designing online communities

Colin Lankshear and Michele Knobel
General Editors

Vol. 72

The New Literacies and Digital Epistemologies series
is part of the Peter Lang Education list.
Every volume is peer reviewed and meets
the highest quality standards for content and production.

PETER LANG
New York • Bern • Frankfurt • Berlin
Brussels • Vienna • Oxford • Warsaw

TREVOR OWENS

designing online communities

How Designers, Developers,
Community Managers, and Software
Structure Discourse and Knowledge
Production on the Web

PETER LANG
New York • Bern • Frankfurt • Berlin
Brussels • Vienna • Oxford • Warsaw

Library of Congress Cataloging-in-Publication Data
Owens, Trevor.
Designing online communities: how designers, developers,
community managers, and software structure discourse
and knowledge production on the Web / Trevor Owens.
pages cm. — (New literacies and digital epistemologies; vol. 72)
Includes bibliographical references and index.
1. Online social networks. 2. Social media. I. Title.
HM742.O94   302.30285—dc23   2014043811
ISBN 978-1-4331-2847-9 (hardcover)
ISBN 978-1-4331-2846-2 (paperback)
ISBN 978-1-4539-1502-8 (e-book)
ISSN 1523-9543

Bibliographic information published by **Die Deutsche Nationalbibliothek**.
Die Deutsche Nationalbibliothek lists this publication in the "Deutsche
Nationalbibliografie"; detailed bibliographic data are available
on the Internet at http://dnb.d-nb.de/.

The paper in this book meets the guidelines for permanence and durability
of the Committee on Production Guidelines for Book Longevity
of the Council of Library Resources.

© 2015 Peter Lang Publishing, Inc., New York
29 Broadway, 18th floor, New York, NY 10006
www.peterlang.com

Printed in the United States of America

CONTENTS

Foreword: Social Sciences of the Artificial   vii

Acknowledgments   ix

Introduction   1

Chapter 1. Learning and Collective Intelligence on the Web   7

Chapter 2. A Framework for Studying Online Community Software   15

Chapter 3. Research Questions and Methods   29

Chapter 4. Community and Values: A Worked Example of Analysis   49

Chapter 5. Rhetorics of Online Community: A Brief History   55

Chapter 6. Enacting Control, Granting Permissions   91

Chapter 7. Studying the Records of Online Communities   119

Appendix: Example Data Collection Sheet   131

References   133

Index   137

FOREWORD: SOCIAL SCIENCES OF THE ARTIFICIAL

Kimberly Sheridan
Associate Professor of Educational Psychology,
College of Education and Human Development &
College of Visual and Performing Arts, George Mason University

Herbert Simon's 1961 *Sciences of the Artificial* is generally credited with framing the contemporary conversation on how "design sciences"—such as engineering and computer science—differ from the natural sciences. Simon guided his fellow researchers into consideration of how they need to approach their work in these artificial worlds. In this book, Trevor Owens continues in this vein, exploring how researchers might think about social interaction in online forums in light of the designed software in which it occurs.

One of the tenets Simon establishes in his initial exploration of the distinction between the natural sciences and the science of the artificial is that although natural sciences are focused on what *is*, sciences of the artificial are focused on what *should be*. Bridges, cars, and computers are designed to function in certain ways, and our studies of how they function are impacted by the principles and ideals that guided their design. Delving into the unlikely source of an over 20-year span of technical "how to" books on designing online forums, Owens provides a careful account of how the designers envision and enact their work, creating the tools that have shaped discussion on the web. Trevor documents how the designers of forum software conceptualize users and how this design gets enacted in the technology of control. He demonstrates how the battles over what the Internet means are fought at the level of how and whether to include "vote up/vote down" options and the ability to "like." Forum software designers and developers pose a vision of the kind of forum user they deem ideal (e.g., are you looking for informed, trustworthy, and civil citizens? eager consumers? loyal fans?) and then enact technical features that reinforce behavior closer to this ideal and punish deviation. Through close reading of seemingly benign practical, step-by-step technical books, Trevor shows how even the management of the minutiae on Internet forums can be fraught with ideological positions and bids for power and control. These ideologies are designed into code that is bundled into features and that gets embedded into new forum sites far removed from their origins.

It is all too easy for those of us who use online discussions as data to forget that we are social scientists of the artificial, or at least to be glib about what that means. To be sure, we think about the idiosyncrasies of the online communities we study and how they impact our findings. We—and our methodological texts—consider how anonymity impacts conversation. We are aware that the text we read is generally a vocal minority of a much larger lurking group. We aim to be appropriately circumspect about extrapolating beyond the text at hand. However, we consider less often the vast and diverse ways the technical infrastructure tweaks, prioritizes, reorganizes, deletes, and otherwise often exerts an invisible control over the conversations. In this volume, Trevor makes a compelling case for why we should consider this and provides some conceptual tools with which to begin the work.

ACKNOWLEDGMENTS

This book, which builds on my dissertation research, would not have been possible without the significant support and insights of my dissertation committee. Each of my committee members played a significant role in the design and development of the project. For the four years I worked at the Center for History and New Media, Dan Cohen was both a great boss and a great mentor who helped me refine a lot of my ideas about online community while helping to grow the community of Zotero users. In his digital history course, I was also introduced to much of the new media studies work that, to a large extent, shapes the argument of this book.

Early in the doctoral program at George Mason University I reached out to Kimberly Sheridan about a study I wanted to do on the RPGmakerVX online community. I had seen from her CV that her dissertation involved Kantian notions of taste and film fan forums. She happily agreed to advise my project, and, under her guidance over the last five years, I have had the opportunity to deepen and refine my critical skills at research design and analytic interpretation. When I first read Joe Maxwell's book *Qualitative Research Design: An Interactive Approach*, I was hooked. It was and remains rare and refreshing to find such clear, focused, and accessible academic writing. Through his writings and experiences in courses, Joe's approach to research has stuck with me as one of my key take-aways from the doctoral program.

Aside from my dissertation committee, many others have played critical roles in refining and developing the ideas and approach in this book. Kurt Squire's ideas about studying forums for the game *Civilization* sparked much of my interest in the topic. Discussions with Ben Devane about discourse analysis of online communities have been invaluable. Historian of science Richard Staley, my undergraduate thesis advisor, who spent far more time than I imagine any other undergraduate thesis advisor has before or since advising a student, was instrumental in setting me up with the habits of writing and picking apart texts that I have made ample use of here. John Levi Martin's seminar on culture and cognition provided me with a significant part of my sociological perspective. More recently, Matthew Kirshenbaum's approach to theorizing and studying software helped to firm up my own thinking. My colleagues working on digital preservation at the Library of Congress have been similarly instrumental in that area. Thanks to my mother, who persisted in convincing me that, despite the expense, I should go to college instead of just hanging out in Milwaukee with the band.

All of that aside, it's most important that I acknowledge my wife and constant collaborator Marjee Chmiel. My very first foray into academic writing and the study of online communities was a conference paper we wrote together about creationist teen web forums. I believe it was Marjee who first spotted the listing for the Zotero job at CHNM and sent it to me so that I could apply. Talking with Marjee about her work on *World of Warcraft* forums helped spark my interest in studying the underlaying software behind online communities. When it came time to work on our PhDs, we tackled them together.

INTRODUCTION

Discussion on the web is mediated through layers of software and protocols. As scholars increasingly study communication and learning on the web, it is essential to consider how site administrators, programmers, and designers create interfaces and enable functionality. The managers, administrators, and designers of online communities can turn to more than 20 years of technical books for guidance on how to design online communities toward particular objectives. Through analysis of this "how-to" literature, this book explores the discourse of design and configuration that partially structures online communities and, now, social networks.

Tracking the history of notions of community in these books suggests the emergence of a logic of permission and control. In these books, the idea of "online community" defies many of our conventional notions of community. Throughout these books, participants in online communities are increasingly treated as "users" or even as commodities to be used. Through consideration of the particular tactics of site administrators, programmers, and designers, this book offers suggestions for how sociologists, anthropologists, folklorists, historians, and other scholars interested in studying the records of online communities should approach the analysis of those records.

Contribution to Understanding Learning Online

Over the past 15 years there has been a proliferation of research on learning and discourse in online communities. In educational research, much of the work on new media literacies is grounded in ethnographic approaches to studying community, cultures, and affinity spaces that have emerged on the web. This book provides insight into the ideology and values behind the software that enables these kinds of online communities and informal learning spaces. As such, the book contributes to ongoing work on online communities in two primary ways: first, it provides a framework and context for researchers studying the textual records; second, as educators increasingly explore using these systems as platforms for creating their own online learning communities, it will give them the necessary understanding of the values and ideology that inform the design and development of the software they are considering.

More and more educational researchers, social scientists, and humanities scholars are becoming interested in studying online communities. This book will fill a gap in existing work on methodology. As such it will be of broad interest as a text or a supplemental textbook for social science courses focused on studying the web. Further, as a study of the values and ideology behind online community software, the book is also valuable to those interested in social studies of the web more generally. Aside from these academic audiences, librarians, archivists, and curators who collect and preserve these records will find relevance in the analysis of records of online communities. Finally, practitioners who design and build these communities can use the review of the history of ideas about running online communities to inform their reflective practice.

Records of Online Communities as Primary Sources

While working on the manuscript, I would occasionally jokingly describe this project as "Reading American Web Forums," a play on Allen Trachtenberg's influential book *Reading American Photographs*. In his book, Trachtenberg documents and explores the nature of photographs as historical evidence. Different from historiography, which studies the history of writing about a particular historical period, Trachtenberg's book is more akin to work in the German tradition of source criticism. Source criticism explores the inherent characteristics, nature, and structure of various primary sources and the processes by which they are created and managed in order to explicate what sorts of arguments one can make based on sources of that kind. In the case of photographs, Trachtenberg shows that many of the assumptions we make about photographs as transparent transmitters of visual information are simply wrong. Photographs are staged. Photographs are composed. Ideas of acceptable practice have changed over time and result in different kinds of photographs. The physical processes of photography have changed and developed over time, resulting in the possibility of different kinds of compositions. In short, to read a photograph correctly as historical evidence, one needs to understand a good bit about how and why it was created and its context in the history of photography. Ultimately, this book makes a similar set of claims about the records of online communities.

I am convinced that, in order to adequately parse the records of an online community, one needs some understanding of the context, platforms, process, mechanisms, and decisions by different actors that result in the records of the community being made available to you. This book can be read as an attempt to develop and refine a bit of source criticism for the study and analysis of this genre of primary sources.

Structure of the Book

I have broken the text into seven chapters. Each chapter builds on the work of the previous chapter. Because it's likely that different audiences may be more interested in particular sections of the book, I will briefly outline the structure of the book here. To enable jumping around to different sections in the book I have composed chapters in such a way that they largely stand on their own. For those interested in getting to the heart of the project, feel free to go straight to the last three chapters, which focus on analysis and interpretation.

Chapter 1: Learning and Collective Intelligence on the Web: This chapter introduces the ideas of collective intelligence and knowledge bases and situates the book in relation to work on informal learning in online communities and new media studies approaches. The goal is to explain why a book like this is necessary. The chapter is relatively brief and serves as an introduction to the issues at hand in the text that follows. Most readers would do well to use this chapter as a point of entry.

Chapter 2: A Framework for Studying Online Community Software: Bringing together work from psychology, software studies, and science and technology studies, this chapter presents an underlying framework for studying the software that enables online communities to provide a robust and substantive synthesis of work from these different fields. This section is likely to be of most interest to researchers and students interested in the intellectual and theoretical foundation for the study of software.

Chapter 3: Research Questions and Methods: This chapter briefly describes the specific methods, approaches, and research questions that the subsequent chapters explore and focuses on explaining exactly how the study was done, why it was done that way, and the inherent limitations of the approach taken. This is of particular use to researchers for two reasons. First, it provides a context for best evaluating the analysis and results in light of exactly how the study was undertaken. Second, it provides a model for how future studies might work through the issues involved in using guidebooks as primary sources for unpacking the ideological functions of software.

Chapter 4: Community and Values: A Worked Example of Analysis: The first of the analytic chapters focuses on a close reading of one paragraph from *Invision Power Board 2: A User Guide*, by David Mytton (2005), which defines the roles and values of online communities. This chapter also serves as a model for how to use Jim Gee's approach to discourse analysis on technical books (Gee, 2005) and as a point of entry to some of the key themes that emerge from the more extensive analytic chapters that follow. As a detailed, worked example of analytic technique, it is particularly useful for those interested in applying discourse analysis techniques to such texts.

Chapter 5: Rhetorics of Online Community: A Brief History: Starting from bulletin board systems of the late 1970s, this chapter explores the vision, ideology, and rhetoric of online communities through the early web, the development of commonly used discussion platforms such as phpBB, and the creation of social networks. It develops and presents the historical narrative of how the ideology and values in the software design have evolved, as evidenced in books about setting up and running an online community. Those interested in studying records from online communities in different periods will find this useful. Understanding changes in the values and ideology behind online communities over time provides a useful context to reevaluate contemporary assumptions about online communities. This historical chapter also provides a basis for understanding the range of competing values at play in current manifestations of these communities.

Those who create online communities or are interested in creating them may find this to be a useful backstory for exploring how their ideas fit into the broader historical context.

Chapter 6: Enacting Control, Granting Permissions: This chapter explores how the creators and managers of online communities use visual design information architecture tools for moderating, filtering, and banning users and reputation systems to push users behave as they would like them to in their communities. This chapter presents a model for the various ways in which these individuals and the software they use structure and shape participation in online communities. Understanding the components of these software systems and the tactics available to administrators to grant permissions and exercise control is invaluable for researchers interested in making sense of the records online. In this vein, the details in this chapter suggest what features of online communities and traces of their processes one should be looking for to inform research based on records of online communities. The methods, tactics, and tools used to establish control in these communities are also of interest to those thinking about designing and running these kinds of systems.

Chapter 7: Studying the Records of Online Communities: This chapter unpacks the implications of the study for researchers and for educators interested in studying, designing and developing online communities. Returning to many of the issues explored in the first and second chapters, I offer a revised set of ideas about how to think about the character of social interaction in online communities. I make specific suggestions for approaches and questions that will help researchers studying records from online communities avoid the significant inherent limitations that come from interpreting these records. The core idea is to approach records of online communities as filtered and selected traces of discourse and not as transparent presentations of conversation and discussion. Given the growing interest of educators in creating and designing their own online communities, I offer suggestions for how educators can become aware of and avoid many of the pitfalls associated with simply picking up features and functionality created with a particular behaviorist and highly commercialized vision of community that is at odds with many of the values and perspectives that undergird educational institutions.

· 1 ·

LEARNING AND COLLECTIVE
INTELLIGENCE ON THE WEB

In a 2009 interview for the popular blog ReadWriteWeb, Mark O'Sullivan, the lead developer of the open-source web forum software "Vanilla," was asked if web forums are still relevant (O'Sullivan, 2009). His response offers a point of entry for understanding the importance of software that powers discussion on the web: "Do a Google search for anything. How many of those search results are from discussion forums?" When asked why there were so many, he responded: "It has to do with people having real discussions and giving real answers." When you go looking for answers on the web, there is a good chance you will find them in some earlier answers to questions.

Users of the web consult this collective knowledge base of questions and answers on a regular basis, but there is relatively little scholarship exploring the structures and systems that create it. In particular, we know little about the design decisions behind the discussion board software and the commenting systems that enable the communications we rely on. Aside from understanding this knowledge base, knowing more about these design decisions can pay dividends for studies of online community and social interaction. The first step in the process involves the software tools that enable and shape our discourse online.

The structure of the conversations that web users engage in on online discussion boards, blogs, and other content-driven platforms is shaped by the Mark O'Sullivans of the world who create, design, implement, and hack on the software that makes the web a platform for community discussion, deliberation, and dialogue. I document and explore the ideology and practices of developers and individuals who implement and configure the software that supports online communities. This analysis suggests how the software and visions of community embedded in that software affect the nature of the discourse in online communities.

Online community sites—first and foremost web forums—are increasingly being explored ethnographically as contexts wherein young and old alike are developing valuable skills and sharing and building knowledge. We learn to write, to create art, to give and receive criticism, and to acquire a range of other skills and knowledge in these online spaces (Ito, 2009). Through extensive ethnographic fieldwork, Ito and her fellow researchers found that significant numbers of young people, supported by web forum software, are "learning to navigate esoteric domains of knowledge and practice and participating in communities that traffic in these forms of expertise" (Ito, 2009, p. 28). Work on fan fiction forums suggests that participants are developing as writers and in some cases using these communities to learn English as a second language (Black, 2005, 2008). Studies of videogame fan forums (Duncan, 2010; Owens, 2010; Squire & Giovanetto, 2008) suggest that participants are developing their abilities to interact with, critique, and design video games.

More broadly, virtual community sites and spaces are being explored as contexts in which civic engagement and democratic practice are developing (Song, 2009). As Song suggests, "virtual communities have captured the public imagination and subsequently become a vibrant site of competing views of 'community' and 'democracy'" (p. 5). As psychologists, sociologists, anthropologists, educators, and others begin to look at online communities, and as those communities come to play an increasingly important role in how people learn and develop knowledge, it is critical that social scientists understand the ideas and theories of community and human motivation that inform how online communities are built and managed. This is all the more important as ethnographic methods are deployed to study situations in which social interaction is entirely mediated by the work of those designing interfaces and modes of interaction. It's not enough to study the experience that takes place on the screen; it's necessary to understand how those experiences result from the interplay of design and the software and hardware that makes what a user experiences on the screen happen.

Interest in the study of online communities has been robust enough that there is now more than a decade's worth of research methods scholarship focused directly on that subject. Methodologies like "virtual ethnography" (Hine, 2000) and "netnography" (Kozinets, 2010) have been developed to translate the ethnographic study of culture and community into the computer-mediated contexts of online communities. At this point there are clearly large and thriving communities, many supported, organized, and sustained through platforms like web forums, and this body of methodological work suggests potent ways of engaging in computer-mediated participant observation and examination of the lived experience of these communities. However, the methodological work in netnography has been primarily concerned with exploring what it's like to participate in an online community rather than how communities are being structured and designed by developers, designers, and community managers within the limits and constraints of software and underlying technical platforms and protocols.

Although there is a growing body of literature on the lived experience of informal learning and interaction in online communities and the textual record of interactions in places like web forums, there remain significant methodological problems associated with conducting ethnographic research in clearly designed virtual environments. One of the primary goals of this book is to enrich the methodological discussion of how to do a qualitative study of these kinds of web communities. It is important to understand the ideologies of the web that operate in both social research on it and in the practices and tactics of the individuals who create and manage online communities.

Locating Power, Control, and Autonomy in Collective Intelligence

In *Collective Intelligence: Mankind's Emerging World in Cyberspace* (1997), Pierre Levy described his vision of the kinds of changes the Internet was bringing to culture. Levy's ideas have "a form of universally distributed intelligence, constantly enhanced, coordinated in real-time, and resulting in the effective mobilization of skills." At the heart of Levy's approach—and that of many other boosters of the web—is the idea that the web empowers individuals and enables the creation of new community networks. In educational

research, Levy's ideas have found particular purchase through Jenkins's artic-
ulation of the participatory culture of the web (Jenkins, Purushotma, Weigel,
Clinton, & Robison, 2009). From this perspective, the web enhances users'
autonomy both by providing access to knowledge and enabling them to create
knowledge. In Levy's words, "The distinctions between authors and readers,
producers and spectators, creators and interpreters will blend to form a reading-writing continuum, which will extend from machine and network designers to the ultimate recipient, each helping to sustain the activity of others"
(Levy, 1997, p. 121). From Levy's perspective, collective intelligence works
to the betterment of all individuals in that "the basis and goal of collective
intelligence is the mutual recognition and enrichment of individuals" (Levy,
1997, p. 39). Much of the work of exploring informal learning in online environments is grounded in Levy's perspective. However, the web is not a free-
for-all. It is an emergent phenomenon, designed and structured by particular
individuals and their software.

    At the most fundamental level, the designs of the web operate on a set
of protocols. The TCP/IP (Telecommunications Control Protocol/ Internet
Protocol) and the DNS (Domain Names System) are the fundamental protocols that enable the web to establish structures of control. As Alexander
Galloway (2006) suggests, these protocols vary greatly in the degree of control they exert over the individuals that use the web. As he explains, "One
protocol (TCP/IP) radically distributes control into autonomous agents, the
other (The DNS) rigidly organizes control into a tree-like decentralized database" (Galloway, 2006, p. 53). At the baseline level, the World Wide Web is
constrained and structured by design decisions in these protocols. Similarly,
Wendy Chun (2005) argues that the structure and design of the communications networks that undergird the web are actually enacted through powerful
control and surveillance of an individual's actions online. The web was "sold
as a tool of freedom," but it can also be understood as a "dark machine of control" (Chun, 2005, p. 2). In Chun's consideration of TCP/IP, this fundamental
protocol establishes conditions whereby "users are used as they use" (p. 21).
As Chun documents, every HTTP request is signed with a user's IP address,
making users' actions visible in HTTP logs even when they think they are
anonymous. Many continue to see the web as a platform that emancipates
users to create and share their ideas and build collective knowledge and intelligence, but that system is highly structured and designed to record and track
users in the way the system's most basic protocols function.

Understanding the protocols of the web is valuable, but it is important to remember that the Internet is not a single monolithic, technological platform but a platform upon which platforms are built. As anthropologists Daniel Miller and Don Slater suggest, researchers need to "disaggregate" the Internet—that is, not to look at a monolithic medium called "the Internet," but rather at a range of practices, software, and hardware technologies, modes of representation and interaction that may or may not be interrelated through participants, machines, or programs (Miller & Slater, 2001). When we sign up for a web forum and begin a discussion thread, when we post a comment on a blog, when we find the answer to a question in a technical forum as the result of a Google search, we are getting something we want, but we are also participating in a designed experience created for particular purposes. For the designers or managers of a particular site or piece of software, that purpose might be to foster "real discussion," as Mark O'Sullivan does, or it might be to maximize web traffic to increase traffic to online advertisers. In any event, exploring and understanding the ends for which different platforms and communities are designed and maintained and the tactics and practices that work to achieve those ends open up the possibilities for understanding the ideologies at play in structuring and constructing the experience of online community.

Considerable research has been done on how users experience and participate in online communities, but little scholarly work has focused on understanding how the people who set up, design, and configure communication on the web think about and theorize their work. Realizing that researchers need to disaggregate the web into a set of distinct platforms and systems that are enabled over the foundational protocols that enable it requires us to think about the ideas and perspectives that inform how particular features, tactics, designs, and configurations are enabled to create particular kinds of results. What tactics do these designers, developers, and community managers use to create the kinds of online communities they want to run? What values, ideologies, and theories of their users and of community itself are reflected in those tactics? What divisions exist within these ideas and ideologies, and how have they developed since the era of thousands of bulletin board systems in the 1980s to the early days of the web and our contemporary world of social networks? These are some of the questions this book begins to answer.

The designers, developers, community managers, and system operators who build, set up, and configure the systems and norms of online communities have produced a technical literature about how and why systems should be built. Books such as *Growing and Maintaining a Successful BBS: A Sysop's*

Handbook (Bryant, 1995), *Hosting Web Communities: Building Relationships, Increasing Customer Loyalty, and Maintaining a Competitive Edge* (Figallo, 1998), and *Building Web Reputation Systems: Ratings, Reviews & Karma to Keep Your Community Healthy* (Farmer & Glass, 2010) have been consulted by the developers and designers behind online communities since the early days of these systems. In many cases, even their titles suggest particular visions for the purpose of community. Are communities things to be grown? Are they places where you can increase the loyalty of customers? What does it say about online community that there are now entire books devoted to the design of technical structures for "reputation systems"? Alongside these general books, guides for particular pieces of software—such as *vBulletin: A Users Guide: Configure, Manage and Maintain Your Own vBulletin Discussion Forum* (Kingsley-Hughes & Kingsley-Hughes, 2006)—offer information about *why* one would run such a system. These texts outline a set of ways of seeing the web, and they document the tactics and practices that guide how web forums and online community sites are set up. For example, here is how Derek Powazek (2002) explained the role of software tools in the introduction to his *Design for Community: The Art of Connecting Real People in Virtual Places:*

> This is all about power. Giving your users tools to communicate is giving them the power. But we're not talking about all the tools they could possibly want. We're talking about carefully crafted experiences, conservatively proportioned for maximum impact. (Powazek, 2002, p. xxii)

Although studies of discourse often turn to explanations based on power and control, power and control are often not described and presented so explicitly by the individuals that researchers study. There is an important tension between the first two sentences of the preceding quote and the last two. In the first two sentences, Powazek is focused on empowering users. The "tools to communicate" are about empowering users, about handing over control. A central theme in these texts is that old media were controlled experiences in which producers produced and consumers consumed. On the other hand, Powazek addresses the soft power of the designer. The act of control comes by way of deciding what tools to provide to a site's theoretical potential users and how one will allow them to communicate. What is important for Powazek is that this is explicitly not about "all the tools they could possibly want"— that is, empowering users is not an attempt to give them everything they want. The designer creates "carefully crafted experiences." The experience of participating in his online community has been explicitly designed toward particular ends. In designing the structure and functionality of an online community for "maximum impact," maximization implies enabling specific kinds of communication between particular kinds of users. Even in his introduction, Powazek expresses a tension between empowering and giving users a voice versus manipulating and restricting users' autonomy.

Software, like any technology, does not exist in a vacuum. It is created, deployed, and managed by individuals and organizations toward their own set of goals. The managers, administrators, and designers of online communities can turn to more than 20 years' worth of technical books for guidance on how to design and structure online communities toward particular objectives. My subject of analysis here is popular how-to guides on running and managing this software. Created by designers and developers themselves as guides for each other, these books offer a window onto the technical and practical, the everyday of designing and managing an online community. How-to guides offer a starting point for understanding the theories of users, of design, and the values that are prevalent in an ongoing discourse about what this software should do. In this sense, I can enter into this discussion at the same point that someone interested in running such a site would and explore the layers of values evident in these texts.

Indirectly, this is a study about how these designers, developers, and administrators have made use of the affordances of the World Wide Web as a platform for enabling the creation of online communities. Working within the constraints of the functionality of the web and with their ideas about what users want and what the community should be, they have developed and refined the software that defines and structures online communities.

This study yields two primary conclusions. First, through a historical analysis of how online community is defined and envisioned in the texts analyzed, I argue that, increasingly, community has less and less to do with the development of social ties and the exchange of knowledge between self-identified members and more and more to do with the concept of an online community itself as a kind of property owned by the site administrator/manager. This finding is useful for ongoing discussions of the extent to which the web is empowering individuals or acting as a system of control. Second, this perspective on the nature of the online community has significant implications for social scientists interested in studying the records of online communities. By understanding the tactics that the administrators, managers, and designers use to shape online discourse, social scientists and humanities scholars can begin to pay attention to many of the impacts of this control on the resulting records.

· 2 ·

A FRAMEWORK FOR STUDYING
ONLINE COMMUNITY SOFTWARE

In 2006 I was working as the "technology evangelist" for the Zotero open-source software project. I was hired to do outreach and help manage the burgeoning online community of users and developers. When the Zotero project launched, its website consisted of a blog, a documentation wiki, and a web forum for discussion. Over the course of the years I worked on the project, I generally spent a few hours a day responding to questions and comments on the project's forums. The genesis of this book traces to an early attempt to introduce some changes to Zotero's web forums. Like many open-source projects, the web forums played a key role in how users got involved in the project. Any user could visit the forums to ask for help, and the answers to their questions served as a knowledge base so that future users with similar problems could search the forums and find the answers without having to re-ask the questions. The forums also served as a platform for users to suggest new features and refine their ideas about how exactly those features would work. An important part of my job was to try to help users troubleshoot issues and to encourage them to become more involved in the project.

At one point I explored adding a plug-in (a software component that adds additional functionality) to the web forums. As with most projects, we did not create our own forum software. Zotero's website uses an open-source web forum software package called Vanilla. I thought it would be useful if users of the Zotero forums could see the user's post count (the number of times the user posted) alongside each of the posts in the forums. In addition, I was interested in setting up rankings for posters based on their post counts—for example, beginner, intermediate, advanced. This sort of feature is found in many web forum systems. The idea behind showing post counts and ranks is that it is easy for new users to see the amount of experience and expertise another user has.

In a search of technical advice on how to add this feature in the Vanilla software support forums, I found heated exchanges on the subject of whether someone might want to add this kind of functionality to his or her Vanilla forums. In one case, someone considering using the software noted that his or her users would like "reputation points" and "user titles." This kind of functionality was not well received by the community of Vanilla users and developers. One user suggested that these kinds of reputation tracking were "pointless" and "more trouble than they're worth" and that an "artificial measure of this is also easily gamed by people and so quickly becomes useless." Another user explained that these kinds of features result in attracting "people that post nothing but mindless drivel just to drive their post count up, and to get a new rank." In a discussion about a similar feature request, a user explained: "Like all revolution, Vanilla's biggest problem is simply that it's different. It's also the strongest feature." I thought Vanilla was just a simple software application that enabled the Zotero team to host discussions, but my ideas about features were in conflict with the "revolutionary vision" that the developers and core users of the software adhered to. Vanilla's vision was one of simplicity—"focus on discussions rather than pointless features." One user went on to explain that these kinds of features "would be going against the Vanilla Movement."

What I thought was a very simple idea about how to modify our online community revealed passionate philosophical disagreements among software designers. This was not something I had expected to find, and the points they raised were issues I hadn't considered previously. I saw my ideas of showing the number of posts a user had made setting up a method of assigning titles to people based on their posts as simple, fun ways to see individuals' involvement in the community. However, I found that those ideas were at odds with the ideology of the software. Something relatively simple to implement from a technical standpoint was accepted because this seemingly minor feature was not in line with the values of those who had worked to design the platform, the Vanilla web forums server side software package. Responses to these rela‑ tively trivial technical requests for software features point to ideological and value‑driven notions of participation and motivation embedded in different software platforms.

These ideologies and values were not simply afterthoughts for the soft‑ ware designers. The individuals working on this software had rather extensive theories about human behavior and social interaction that informed their ar‑ guments about design. In my particular example, software designers were op‑ posed to features that led people to "game" the system by suggesting that the value of participation in the community was captured in a point system. They had a vision of what an online community and discussion should look like and how it should work. Those ideas were part of an ongoing technical discussion about the design of these systems.

Theorizing Online Community Software

As Nick Montfort and Ian Bogost (2009) suggest, the study of software involves the study of layers of software on top of software intertwined with particular pieces of hardware (2009). The layers in these platforms provide particular affordances and constraints but are generally taken for granted by users as a part of the platform. The subjects of this book—the server‑side software packages and scripts discussed in the how‑to guides—are a particu‑ lar layer in the stack of software and hardware that enables the existence of online communities.

The software layer implicated in these texts is the software that users are directly interacting with on the server when they participate in online com‑ munity websites. Often this software comes in the form of applications such as phpBB, Vanilla Forums, the Ultimate Bulletin Board System, or vBulletin. In other situations it is made up of custom PHP, Python, or Perl scripts that interact with server‑side databases. While we experience the web as a series of individual pages that load one at a time in our web browsers, and while many of us understand that web pages can be individually written in HTML, most of our interactions with the web are actually mediated by these other software packages. The above‑named software systems and their databases establish rules for what users can and cannot do, what users can and cannot see, and how others in an online community site see users and their online actions. In short, my object of analysis is the layer of software that exists just below the surface (from the user's point of view) and on the servers that we interact with via HTTP.

Most studies of online communities analyze and interpret the user experience—that is, they interpret the rendered pages that are created and displayed on the user's screen. The subject of analysis is the ideas behind the software that runs on the server, in other words, the software that an administrator runs and configures that enables the interactions that end users experience. Where one looks to understand how users interpret rendered pages, it is important to understand how developers and administrators understand and think about the software on the server, since their ideas about this software have a clear impact on the experience of users. In this respect I am indirectly studying software and directly studying how this software is designed, deployed, tweaked, and made to create user experience. My approach is to consider the interplay between these technical mechanisms and the textual discourse in which the norms, values, and ideologies that administrators and developers bring to their work on these systems are articulated. By helping researchers to further disaggregate the Internet and to appreciate the roles that developers, designers, managers, and software play in structuring participation in online communities, this approach is intended to inform methodologies for studying participation in online communities.

Theorizing Software from Technology Studies

In cultural studies scholar Matthew Fuller's *Behind the Blip: Essays on the Culture of Software* (2003), it is argued that "each piece of software constructs a way of seeing, knowing, and doing in the world that at once contain a model of part of the world it ostensibly pertains to and that also shape it every time it is used" (Fuller, 2003, p. 19). Through the analysis of a variety of software platforms, he argues for interpretive study of how software constructs these ways of seeing. It is critical for researchers of online learning to realize that software operationalizes a set of ideas in a designed experience that constrains individuals' experiences, participation, and learning. Software shapes much of our experience; however, it is too simple to say that software itself constructs "a way of seeing, knowing and doing." Software affords and suggests particular uses to users; it shapes and structures experience. The meaning and utility of software for people (in this case both those running and administrating the software and end users of sites powered by the software) always involves meaning making between the person and the artifact. If we consider insights gleaned from studies of other technologies, the meaning of any technology, like software, has a complex cultural and symbolic process of definition.

Users matter in software systems, and, given the extensive body of historical work that has focused on understanding the relationships between users and designers of technologies, it is important to situate analysis in work on the history and social study of technology. In the 1980s, historians of technology Trevor Pinch and Wiebe Bijker (1984) drew attention to the users of any given technology through an approach they called the social construction of technology. The key concept in this work was that different individual users were able to construct radically different ideas about what a specific technology could or should be used for. Bijker made the relationship between users and designers of a particular technology more explicit with his idea of the technological frame. In this conception, the users and designers are thought to negotiate and then agree to a particular interpretive frame for understanding the use and value of a given technology. Stephen Woolgar argues that we should approach technological artifacts as texts where the designers are authors who are actively involved in "encoding" particular meanings and uses into technologies that the individual user reads and interprets (Woolgar, 1991). With this noted, focusing on "encoding" can yet again lead to ceding too much authorial intent in the process of reading these machines as text. As Madeleine Akrich suggests, users are also involved in a process of "decoding" that text. The relationship between users and designers has also been conceived as similar to that of a film script: "Like a film script, technical objects define a framework of action together with the actors and the space in which they are supposed to act" (Akrich, 1992, p. 208). The lesson from perspectives in the history of technology is that much of what a technology does and means is the result of an interpretive process by which the meaning, implications, and rules for the use of a given technology are negotiated between a range of users and designers. The guidebooks for online community software offer insight into how the creators and administrators of online communities anticipate and negotiate their ideas of users with the design of their systems.

Bruno Latour's Actor Network Theory (ANT) brings together ideas about how to interpret software suggested thus far (2005). In Latour's approach we understand the social world as a network of interactions between actors, both people and objects. In this case, things like how-to guidebooks are themselves tools that exist as part of those systems. Broadly speaking, I would suggest that Actor Network Theory allows researchers to retain materiality of software.

Instead of focusing only on what actors do, Actor Network Theory introduces the idea of "actants," that is, anything that modifies or acts on something else. Unlike an actor, actants are not tied up in questions about agency. As a material actant in the network of action, any given piece of software enables, disables, or suggests particular actions for its users. The phpBB web forum system's default settings suggest particular ways to set up a given online community, and the structure of its design limits what a particular community manager can and cannot change about the site they create and manage with the software. From there, the site that the community manager creates and manages is itself a material artifact that, as a result of decisions made in setting up and configuring the software, enables, disables, or suggests particular uses for the participants in the online community. Because of the platform nature of the web—that is, layers of software set atop layers of software—this kind of regress between users and software can be mapped back even further. The designers of the phpBB software are constrained by the features of the PHP scripting language and the protocols that undergird the web (TCP/IP and the DNS). Again, each of those software components (the PHP language, TCP/IP, and the DNS) have been designed by particular people with their own ideas and visions, and thus the network of actants encompasses an extended regress back through people's ideas about software, which are codified into software products that create particular constraints that prompt users of software (both people and other software) to create additional platform layers.

The material nature of software as an enabler, disabler, and suggester is still always contingent on individual uses of discourses that include rules and norms for social interaction around that software. It's not that everything is socially constructed. Importantly, particular software and systems enable or disable particular kinds of use. However, the affordances of those systems are used and interpreted through individuals cultural frames and scripts. In this context, examining how-to guides offers a point of entry into the discursive tool-kit (the ideas about how people and software should act, shared and contested ideas about what is and isn't ethical based on particular values, and so on) that those creating online communities are working with and from. This discourse is the story lines and cultural script in these texts. You can see it in stories about miserable users and what you need to do to keep them from ruining your online community, in the values on display in reports on how to set up your comment systems to get "quality discussion," and in the values that drive particular open-source software projects about "openness." In the context of Actor Network Theory, this discourse is itself an important node in the network of interaction among users, software, servers, designers, and administrators. These storylines, themes, and values play a causal role in shaping how the designers and administrators of software respond to their users and design and structure their software.

By focusing on the how-to guides, I gain access to some of the places where theories and values can be interrogated on the page. I can unpack perspectives on users and functionality in these software systems. The discourse, the cultural models and scripts, around the network of actors (users, designers, administrators, and so forth) play an important role in our understanding of them, but so do the physical and material properties of the software—protocols like HTTP and physical objects like servers and cable. The text of the how-to guides represents attempts to organize and make sense of this network of people and things. The social and cultural is not to be understood as some outside force; in Actor Network Theory, social forces are more accurately thought of in terms of internalized theories about others. In this case, how-to books are interesting as each book offers access to the theories of users and user behavior that have been committed to the page and disseminated as cultural scripts for other developers and administrators to look up and potentially integrate into their internal theories of users. To be sure, like all cultural notions, these ideas are neither uniform nor universally shared. Instead, each text offers a point of entry into the ongoing discursive activity to define and make legible the roles of people and technologies as actors in the network of action.

Collectively, these approaches to thinking about the social construction of technology, of the ways authors encode meaning into technology, and the way that technologies' roles in society play out according to scripts draw attention to the important role that the texts around a given technology, such as how-to guides, play in establishing the functionality and role of a given technology. The give and take between these perspectives suggests that the meaning of particular technologies and tactics is not fixed and deterministic, but instead is negotiated and argued for. For instance, when working on the Zotero forums, the idea of reputation points—displaying a running numerical score for a user's contributions—initially seemed like a simple and straightforward idea for getting more users involved. However, in exploring discussions about this feature in the Vanilla software forums, I quickly found out how technical discussion of an individual feature was fraught with argument about human motivation and the values of community in the underlying software product that was used to run the Zotero forums. This suggests the value in looking at texts written by different kinds of authors—for example, texts from those interested in making money online vs. those of individuals invested in causes such as open-source software. The meaning of particular approaches and functionality are likely to be contested and to be framed and approached by different authors from different perspectives. Explicitly seeking out this diversity creates a richer set of approaches to thinking about online community software.

By exploring how these guidebook texts frame, present, and suggest the value of particular elements of the technical components of software, I investigate the interplay between discourse and technology. In these books I have found that the authors offer up a significant number of stories and anecdotes drawn from their experiences to contextualize their advice on how to design and configure online communities and online community software. (Instead of being entirely prescriptive, these texts often wrap their recommendations in stories about particular kinds of theoretical users.) I'm curious about how the cultural models and scripts evident in these texts and in their configuration stories—stories of configuring software systems and stories of configuring online communication and discourse itself—suggest a range of valuable lessons for understanding the nature of online discourse. Throughout my analysis, I work to suggest how researchers can use an awareness of how online community is itself configured to make better judgments about what one can infer from the record of conversation that persists in online communities. As studies of informal online learning communities continue to flourish in research on literacy, instructional technology, and educational technology, this kind of baseline understanding of the design of these spaces has significant potential to inform and improve this kind of research.

Cognitive Systems and Cognitive Niches

Beyond thinking about these technologies from the perspectives developed in software studies and science and technology studies, these software platforms function as cognitive systems. In this section I establish the importance of understanding cognition as something larger than an individual, as a distributed property of systems. From there, I suggest the need to think about the technical decisions concerning web forums and other ideas about technical systems for facilitating community interactions on the web as shaping our own personal and social cognitive systems. Technologies co-create our cognitive apparatus. That is, as cognitive activities, learning and education are fundamentally mediated by technologies. To this end, unpacking how software is deployed to create spaces for online communication is important for understanding the cognitive niches and feedback cycles in knowledge production and learning.

How technologies distribute cognition. In "How a Cockpit Remembers Its Speed," cognitive anthropologist Edwin Hutchins (1995a) makes a compelling case for thinking about cognitive systems as something larger than individuals. By recounting in detail all of the information that is processed between the individual pilot and the technology in the cockpit, Hutchins carefully documents the extent to which the system of the pilot, plus the tool, is acting as a cognitive whole. In a much more extensive study of ship navigation, Hutchins documents how the ship as a technological artifact can best be thought of as a cognitive network. In this case, it is not simply an individual and a tool (like the cockpit), but is rather a network of individuals using different components of the ship to enable collective action and decision making (Hutchins, 1995b). Both cases suggest the value of thinking about cognitive systems as something larger than individuals.

Thinking of cognition as something that results from a network of tools and individuals leads to an even more expansive conception of cognition. More broadly, acts of cognition incorporate the intended and unintended results of the ideas of the designers and engineers who create, assemble, and manage our tools of thought. It is not simply that these individuals have created objects to accomplish these goals, but that the work of the designers of these artifacts in creating them can similarly be understood as an element of the cognitive act. The tools we use are not simply instrumental in cognition. They are themselves part of our cognitive systems—part of our expanded minds. From this perspective, studying the design and structure of communication platforms such as web forums and other online communities, which have become part of our everyday thought processes, is also partly a study of the networked cognitive system of learning and knowledge we are collaboratively constructing.

Cognitive niche construction. Technologies co-create our cognitive apparatus. Let's return to Mark O'Sullivan's comments about the role that question-and-answer forums play in helping web users find information. In that case, Google's search algorithms, in combination with these question-and-answer forums, create cognitive niches that we inhabit by creating knowledge feedback cycles. Cognitive philosopher Andy Clark draws attention to the idea of cognitive niche construction, a term he builds off of the evolutionary biology notion of environmental niche construction.

In evolutionary biology, niche construction refers to how organisms, "to varying degrees...choose their own habitats, mates, and resources and construct important components of their local environments such as nests, holes, burrows, paths, webs, dams, and chemical environments" (Clark, 2008, p. 131). In each of these cases, animals' behavior has altered their environment, and those alterations then become the basis for further adaptation. This notion of niche construction fits quite nicely with how I am conceptualizing the relationships between software platforms, users, and designers. The software platforms are cognitively akin to the nests, holes, burrows, paths, webs, and dams that animals create.

One of the primary examples of this is the spider's web. "The existence of the web modifies the sources of natural selection within the spider's selective niche, allowing subsequent selection for web-based forms of camouflage and communication" (Clark, 2008, p. 61). The spider's web is interesting as an example of an individual organism and its tools, but beyond this the example of a beaver's dam brings in far more complexity. Dams are created and inhabited by a group of individual beavers and, further, are extended over time, outliving the individual beavers that occupied them. Further, beavers adapt to the niche that the beavers before them created and the altered physical landscape that that dam has produced. These individual dams outlive the beavers themselves. The dam one generation builds structures and shapes the development of the next generation. What matters for Clark in this case is that "niche-construction activity leads to new feedback cycles" (2008, p. 62).

Technologies co-create our cognitive apparatus: the cockpit, the ship, web forum software, and Google's search algorithms affect our cognitive niches by interacting in our knowledge feedback cycles. The users of a particular piece of software—whether the designer, the engineers, or the person who deploys it—are each nodes in a network of activity and feedback that continues to shape the niche. These technologies, and the ideas behind them, are not just tools that help us learn but are part of our extended cognitive apparatus. As more and more people participate further with these kinds of designed online communities, the impact of the designers' and administrators' ideas and theories of community become integrated into the cognitive infrastructure of knowledge, memory, and thought. This is particularly salient in the somewhat utopian notion of collective intelligence. Where distributed intelligence is primarily descriptive—that is, it explains a general characteristic of intelligence as something distributed throughout systems of people and tools—collective intelligence is largely prescriptive, envisioning and designing roles for people as components in knowledge systems.

Collective Intelligence in Action

These notions of distributed cognition and intelligence fit well with Levy's previously described concept of the World Wide Web as a platform that creates and sustains collective intelligence. What is particularly important about Levy's ideas about collective intelligence is that, unlike ideas of distributed cognition, Levy's theory is being actively used by those who are developing, designing, and creating the structure of online community systems.

In *Programming for Collective Intelligence*, a 2007 technical book targeted at developers and designers that is included in the collection of books I intend to study, author Toby Segaran describes Wikipedia and Google's search algorithms as examples of how to build tools that make use of the collective intelligence offered on the web. The concept of collective intelligence—the idea that the web enables the creation of extensive stores of knowledge that can be used to help filter, sort, and provide information to all of us—is part of how designers and developers envision their work. In this case, the expansive online networks are understood as a kind of extensive external memory. Similarly, collective intelligence includes some of the notion of actor network theory, in which this collective body of knowledge is made usable between various actors in the network.

The structures of the systems that facilitate creation of this collective intelligence are explicitly designed to shape the communication that creates that knowledge. This is not to say that those designs don't come with numerous unintended consequences. Situating this project in work on software studies helps suggest ways of examining software and the process of creating that software as artifacts and texts for analysis. Similarly, studies in science and technology suggest the value of approaching this kind of communications technology as infrastructure, while simultaneously offering a set of cautionary tales for interpreting the relationships between the ideas of what a given technology does and the effects of those ideas. Finally, work on distributed cognition similarly focuses on the way technology is embedded in a network of individuals and tools that enables particular kinds of thinking. Together, these perspectives suggest the value of analyzing and contextualizing software and various texts related to the creation of that software in order both to understand these objects and to begin to see the way these tools are being constructed to achieve particular ideas of online community.

Learning in online communities as participation in collective intelligence. A flurry of educational research has focused on studying the kinds of skills young people are developing in online learning communities. Much of this work has focused on how online learning experiences are helping individuals develop "new literacies" with digital media (Lankshear & Knobel, 2006) as well as the kinds of valuable skills young people are developing in online communities related to their interests in video games. Levy's notion of collective intelligence is similarly operative in these studies of informal learning on the web.

In a study of discussion threads in the *World of Warcraft* forums, Constance Steinkuehler and Sean Duncan (2008) and Steinkuehler and Marjee Chmiel (2006) find that beyond serving as a space for discussion, the threads also serve as a knowledge base. Drawing on Levy's use of the notion of collective intelligence, the authors suggest that the discourse and dialogue between these gamers becomes a body of collective information that is then consulted by others as a resource. In this case, they suggest that the collaborative construction of knowledge in *Warcraft* forums parallels the kind of collaborative construction of knowledge that occurs in scientific communities. By looking at the arguments that players engage in about how to best use resources in the game, Steinkuehler and Duncan document the sophistication of argumentation in this space. Further, Steinkuehler and Duncan suggest that beyond simply documenting discourse, the discussions themselves become resources that other players draw on to make decisions.

We can see another component of Levy's reading-writing continuum in the way many games invite players to modify them. In a study of the game *Civilization III*'s forums, Kurt Squire and Levi Giovanetto (2008) found that participation in the online forums served a similar collective knowledge production role as the *Warcraft* forums. Beyond this, Squire and Giovanetto suggest that the *Civilization* forums scaffold individuals interested in playing the game into developing the ability to modify and redesign the game. Players participating in these forums clearly develop technical skills. Squire and Giovanetto suggest that "More important than the particular facts or technical processes may be the practice of negotiating social organizations (including forming them) to further one's own learning" (Squire & Giovanetto, 2008, p. 27). The conversations in these forums are not simply dialogues; they are an organization of knowledge resulting from a process of surfacing the most important parts of that discussion for others to find and play games, comment on and critique games, and create their own games. Participating in—and making sense of—these kinds of online communities may in itself be an important skill. That is, Squire and Giovanetto are suggesting that learning how to *use* collective intelligence is becoming an important skill.

Online communities don't simply enable a range of constructive and creative learning activity. James Gee and Elisabeth Hayes (2010) have shown how different women and girls involved in writing fan fiction and creating films using video games and other kinds of media developed as designers, gained audiences, and found their voices by using various versions of *The Sims* and participating in online discussion boards as platforms for learning. For Gee and Hayes (2010) there is a stark contrast between the kinds of learning that occurs in these interest-driven online communities and learning in schools. They suggest that schools, "which now stand so separate from the rest of the learning landscape, will have to integrate with other means and locations of learning" (Gee & Hayes, 2010, p. 150). From their perspective, it is this informal network of forums and discussion spaces that has become the primary site of learning in our society.

Together these examples illustrate the various kinds of skills with media production and creation, or what literacy specialists Colin Lankshear and Michele Knobel (2006) call "new literacies." Lankshear and Knobel have encouraged scholars to explore those communities in order to develop innovative ideas for formal learning environments. There is broad interest in studying informal online learning communities as a means of invigorating classroom practice (see, for example, Greenhow, Robelia, & Hughes, 2009).

What I suggest, however, is that it is not enough to study these sites as places where informal learning occurs. As previously discussed, decades of research in the history of technology demonstrates that technology starts a conversation between user and designer. Technologies are theory-laden, designed by people with implicit and explicit ideas about the way the world— as well as other humans—works. Therefore, if we are to understand discussion forums and the informal learning that can blossom in these forums in a profitable way, we need a greater depth of knowledge about how they are designed and structured to produce particular outcomes and goals.

· 3 ·

RESEARCH QUESTIONS AND METHODS

My research questions are specifically framed as questions about the information and perspectives presented in the guidebooks written by discussion forum software designers. The answers to these questions have implications for those interested in studying online communities and those interested in understanding the history and development of ideas and perspectives on features used by many popular online community platforms and systems.

The analysis in this book focuses on three primary research questions:

1. What values and psychological theories of users and social theories of community are evident in how books about online communities describe and present online communities? Further, how do the designers' values and social theories appear to influence the designs, techniques, and approaches they present?
2. What tactics do these authors describe for shaping online discourse, and what values and social theories are implied in those tactics? Specifically:
    a. What information architecture and visual design approaches do they suggest?

   b. What techniques and approaches do they suggest for controlling users, such as moderating discussion or banning accounts?

   c. What implementations do they suggest for setting up reputation systems and user profiles?

  3. What substantive differences concerning these tactics, values, and theories exist in the books? Do these ideas change over time? Do they represent distinct ideological or professional perspectives?

Values and Ideology

Question 1 focuses on the ideological nature of developers' and community managers tactics. The term "online community" has meant different things to different people at different times, and unpacking the evolution of the term suggests some of the values at play in different visions of it. By examining how theories of human behavior and social organization and community are read into or used to argue for particular designs, I draw attention to the ideological nature of these designs and shed light on how particular functionality is connected to particular ways of thinking about community. In this case, I am particularly interested in how autonomy and control are described. Authors of these texts regularly explicate their theories about individual human behavior and motivation and how social hierarchy and community structure interact with individuals.

Tactics

Question 2 investigates how ideas and values are embedded in the particular features and designs that the authors of the books are advocating. While I broadly explore the authors' perspectives, I also focus particular attention on the visual and structural designs, approaches to moderating the discussions, and reputation systems. In my initial study of three books, each of these issues served as particularly rich sets of information for explicating the ideological perspectives of the authors. As a result, I made them primary focal points of my analysis of the entire set of books. Looking systematically at these features in each of the books will help to provide focus to my study. By spotlighting these issues, I underscore how seemingly dry, technical configuration details act as sites in which the more interesting and extensive social components of Question 2 are enacted.

Looking for Difference

Question 3 focuses on exploring the differences in ideology and perspective and enactions in software that exist between the books. I am interested in documenting how the ideas about these software systems have changed in the 22 years in which I have found guidebooks about running online/virtual communities—specifically, starting with books from the pre-web hobbyist era of Bulletin Board Sites (BBSes) and tracing them through to the emergence and development of a thriving web industry. For instance, there are tensions between the free software movement and culture and the culture of business and marketing.

Research Method and Design

This project treats how-to and user guidebooks as participants in a discourse, an ongoing dialog defining online community and the best tactics to use to build and sustain online communities. In this section I briefly explain my theoretical approach to discourse analysis and why I think that approach is valuable for studying these guidebooks. I then describe the books included in my analysis and the process I used to select a diverse set of 28 books published between 1988 and 2010. I then explain my analytic focus and process.

Analyzing Texts as Discourse

I approach analysis of these texts in keeping with Norma Fairclough and James Gee's approaches to the study of discourse (Fairclough, 2003; Gee, 2005). While in many circles the study of discourse is often associated with Foucault's work on power and institutions, Fairclough and Gee are more broadly concerned with connecting that approach with sociolinguistic methods of understanding how we do things with words. In this tradition, part of discourse involves what Gee (and many others) refer to as "cultural models," the storylines and cultural scripts that make sense of our individual situated meanings. In this respect, guidebooks describe a particular functionality of software and explain when that functionality is and is not appropriate for a given set of goals.

This approach to discourse is not focused on revealing social or cultural forces, but rather on how individuals use these texts for "world building," an approach that is based on how people use words and texts to do things. In Gee's words, "People use language to communicate, cooperate, help others, and build things like marriages, reputations, and institutions. They also use it to lie, advantage themselves, harm people, and destroy things like marriages, reputations, and institutions" (Gee, 2005, p. xi). The stories and the explanations of particular tactics and techniques in these books suggest particular goals, values, and ways of thinking about users, and I am interested in thinking about what these words tell us and what these words do.

The value of studying discourse in how-to guides as software studies. Given my interest in online community software and online communities, it might seem a bit strange that I propose to study books. In short, I have found these how-to books to be particularly rich sources not simply for describing particular tactics for designing online communities, but also for explicating how and why one should design and configure online communities toward particular ends. Specifically, these texts present especially thoughtful, robust, and coherent perspectives of developers and administrators. Beyond this, they are also influential texts in their own right.

As published works, these guidebooks represent thoughtful and reflective organizations of knowledge and experience. They present the well-developed working theories of those with technical experience in administrating, developing, and designing this software. These are not typical users' ideas or understandings of these systems. Rather, these are accounts of individuals who have likely spent much more time thinking through and developing a theory of their experience. To be sure, the authors may be more inclined to put together a very coherent account of their beliefs about users and their connections to their tools than the actual messiness of design. In practice, it is likely that they might have a muddled sense of users and make technology decisions on what is easiest or what they happened to think of first. These accounts should not be thought of as specific, accurate presentations of what happened, but instead as presentations of theories that have developed through writing and talking about these sites.

It is worth pointing out that this is not simply a limitation. As John Levi Martin suggests, when social scientists, "[t]he self-appointed auditors of behaviors swoop down upon actors" and ask for explanations of individuals' actions, "it is hardly surprising that actors' retrospective scrambles to put their affairs in order—their stories of their motivations—are often unsatisfying" (Martin, 2011, p. 105). In contrast to "swooping down" on community developers and administrators and interviewing them, or sitting down next to them and looking over their shoulders, it is useful to start with these how-to books, which represent some of these actors' attempts to develop and articulate more thoughtful presentations of their perspectives.

These perspectives are particularly important because, as popular technical books, they have likely influenced the work of designers and managers in the field. That is, they provide access to theories *for* as well as theories *of* the design. As widely available published works, these texts are themselves sources marketed to developers and administrators of online communities for them to read up on how to run online communities. In terms of Gee's discourse analysis, these books play an important part in explicating, defining, and describing the "cultural models" or "cultural scripts" in the discourse of online community design and management. The audience for these books is looking for advice and information about how to design and manage online communities. In this respect, these how-to guides are intended as a resource on which developers and administrators can rely for advice on how to go about their work. The fact that there is a market for these kinds of guides is indicative of the fact that there is an audience for them. This is not to make too much of the role these books play in shaping or defining larger cultural scripts for designing and managing online communities. As the contribution of working practitioners, the content of these guidebooks reflects working theories within the community of software designers and managers. To this end, this book is not about identifying the source of particular ideas or perspectives, but instead about using these books as points of entry to document the values and tactics evident in an ongoing discourse between the designers and managers of online communities.

Studying these texts brings with it another clear benefit. Sociologist John Levi Martin has argued that the myriad problems associated with defining the best explanation of action in social science as third-person causal forces make it a fundamentally flawed approach (Martin, 2011). For too long, the social sciences have attempted to identify "social forces" that cause individuals to do what they do in given situations. In doing so, they have reified statistical constructs into things that act in the world instead of understanding them as aggregates of individuals' actions (Martin, 2011). In many cases, individuals' beliefs and ideas are themselves causal forces that shape the world (Maxwell, 2004). Despite problems associated with the analysis of first-person explanations of action, we can at least say that those first-person explanations are actual things in and of this world. Taking Martin's ideas seriously, the best place to begin our understanding of the social is in a rigorous engagement with actors' first-person explanations of their actions and desires.

By focusing on how these texts frame, present, and suggest the value of particular elements of the technical components of software, we can explore the interplay between discourse and technology. These configuration stories—stories of configuring software systems and stories of configuring discourse itself—suggest a range of valuable lessons for understanding the nature of online discourse. These stories offer focus points for exploring how online community is itself configured. An analysis of these texts can help enable researchers to make better judgments about what can be inferred from the record of conversation that persists in online communities.

Book Selection

This analysis focuses on 28 books published in the period 1988 to 2010. The intended audience for these books is individuals who want to set up, run, design, or build their own online communities. The existence of this audience indicates that there is a market for books about setting up and running online communities, which in itself is a noteworthy point for consideration. By 1988, online communities in the form of Bulletin Board Sites, or BBSes (which will be described later) had become something of interest to enough people to warrant the publication of an array of books. Given the focus of my project, I have not included memoirs about experiences in online communities, academic books about studying or evaluating online communities, or computer science research publications focused on model and system design.

There is significant diversity among the 28 books. They were written by professionals from a variety of backgrounds, including web designers, web developers, community managers, systems operators, open-source software leaders, and business and marketing professionals. Some of the books focus on particular design components such as reputation systems. Some focus on how to use particular software platforms such as phpBB or vBulletin. Some broadly discuss using a range of different social media platforms, like using Facebook or MySpace alongside discussion of mailing lists and web forums. Others focus specifically on web forums and systems for commenting and moderation. In each case, these books include analysis of asynchronous text-based discussion and interaction. The primary focus of this project is on asynchronous text-based interactions and the systems created around those discussions and comments, as well as the systems that shape the resulting discussions.

Identifying books. Given my interest in popular technical books and how-to guides, I started to look for relatively contemporary books through searches on Amazon. I searched for books on topics such as designing online communities, managing web forums, running online communities, and for a range of specific software platforms for web forums (phpBB, vBulletin, Invision Power Board, UBB). By finding these books, as well as books Amazon suggested as "related," I identified half of the targeted books. Amazon's listings cater to contemporary tastes and interests. Given my own interest in studying change over time, I took information about these books and found related books in the Library of Congress catalog. By looking up each of the books, I found that I was able to identify the various subject headings each book had been categorized under. Looking through subjects such as "Electronic villages (Computer networks)"; "Virtual communities"; and "Electronic discussion groups," I was able to identify several more related books going back to the 1980s.

Including BBS books. While the primary focus of this book is on asynchronous discourse on the World Wide Web, I have also decided to include a series of how-to books about running and managing computer-based Bulletin Board Systems (BBSes). From 1978 through the mid-1990s, a range of computer-based Bulletin Board Systems enabled individuals around the world to dial in to a particular system and post and share messages. The development of web forums and web BBSes drew on the functionality and design of earlier Bulletin Board Systems. Thus, books about BBSes are themselves an important part of the discursive history of online communities. The earliest books in the collection (from 1988 to 1994) are primarily books about BBSes. There is a clear conceptual continuity in the structure and design of these books with the later web-focused ones. They similarly describe how to go about configuring software with particular goals in mind, how to attract a set of users, and how to manage and moderate them. Aside from this, many of these books also include mentions of emerging approaches to running their software on the web.

Table 1. Chronological Table of Books by Date, Number of Amazon Reviews (AZ), Number of Copies in WorldCat (WC)[1] Participating Libraries, and Target Audience.

Title	Publisher	Year	AZ	WC	Target Audience
The Complete Electronic Bulletin Board Starter Kit	Bantam	1988	0	47	Community Manager; Designer; Business
Using Computer Bulletin Boards	Management Info	1990	0	168	Business; Community Manager; Designer
Bulletin Board Systems for Business	Wiley	1992	0	105	Business; Designer
Running a Perfect BBS	Que	1994	1	51	Community Manager; Designer
The BBS Construction Kit	Wiley	1994	1	87	Community Manager; Designer
Growing and Maintaining a Successful BBS	Addison-Wesley	1995	0	0	Community Manager
New Community Networks: Wired for Change	ACM Press	1996	1	371	Community Manager
Net Gain: Expanding Markets Through Virtual Communities	Harvard Business	1997	63	3	Business; Marketing
How to Program a Virtual Community	Ziff-Davis	1997	1	38	Designer; Developer; Community Manager
Virtual Communities Companion	Coriolis Group	1997	0	0	Community Manager
Hosting Web Communities	Wiley	1998	6	238	Marketing; Community Manager

Title	Publisher	Year	AZ	WC	Target Audience
Community Building on the Web	Peachpit	2000	25	24	Designer; Community Manager
Online Communities	Wiley	2000	15	400	Designer
Poor Richard's Building Online Communities	Top Floor	2000	5	61	Designer; Community Manager
Design for Community	New Riders	2002	12	215	Designer; Developer; Community Manager
Building Online Communities with phpBB 2	Packt	2005	6	225	Community Manager; Designer; Developer
Invision Power Board 2: A User Guide	Packt	2005	3	192	Community Manager
Building Online Communities	Apress	2005	29	372	Designer; Community Manager
vBulletin: A Users Guide	Packt	2006	1	4	Community Manager
Programming Collective Intelligence	O'Reilly	2007	87	423	Developer; Designer
Managing Online Forums	AMACOM	2008	62	714	Community Manager
The New Community Rules	O'Reilly	2009	32	408	Business; Marketing
Online Communities Handbook	New Riders	2009	2	109	Business; Marketing
The Art of Community	O'Reilly	2009	44	184	Community Manager; Open Source
Building Social Web	O'Reilly	2009	14	243	Designer; Developer
Designing Social Interfaces	Yahoo Press	2009	9	48	Designer; Developer

Title	Publisher	Year	AZ	WC	Target Audience
Design to Thrive: Creating	Morgan Kaufmann	2010	21	539	Designer; Developer
Building Web Reputation Systems	Yahoo Press	2010	8	121	Designer; Developer

[1] WorldCat is a union catalog of books available at libraries around the world. Thus, the number of copies of a book in WorldCat member libraries offers information about a book's availability, success, and longevity.

Analytic Focus

Based on initial exploratory research on a subset of the books, I will concentrate on four specific areas of focus in these texts. For each area of analytic focus, I have provided my reasons for selecting it.

Visual design. This includes explanations of how to use visual design and site structure to achieve particular objectives, as well as using visual design to prompt users to act in specific ways. As these ideas focus on parts of sites that are visually evident, documenting them would be of value for researchers studying online communities, who could then use discussion of visual design in these books to help interpret the online communities they study.

Banning users and moderating user content. This includes discussions of when, how, and why one should ban users and what one should do with their accounts and profiles once they are banned. Discussion of how, when, and why one should or shouldn't moderate user-created content is similarly relevant. I have found this to be a valuable counterpoint to the focus on site and page design. Advice on this topic focuses on things that aren't likely to be evident from looking at a particular community's website. Furthermore, focusing on banning and moderation gets at the heart of my theoretical interest in the relationships among software, underlying protocols, and the kinds of agency that a developer or administrator can exert in relation to the kinds of agency that users and participants in these communities can exert. In this sense, looking at the technical issues related to banning users and moderation of content is helpful for uncovering issues around control and autonomy and the roles that social norms play in determining what kinds of things are and aren't appropriate for developers to consider, given particular patterns of user behavior.

Reputation systems. This is focused on explanations of why one should or should not set up systems that track and show users each other's status regarding how prolific or popular a user's posts are. I am particularly interested in how authors conceptualize users' motivations to contribute to the forum. Given my own experience with the Zotero forums, I believe this is a great place to uncover potentially differing perspectives about user psychology and motivation. By default, designing these types of systems requires conceptualizing "reputation," which quickly pushes authors to suggest models of social value and participant motivation. Given the interest in things such as "badging systems" for online education and other applications of the development of these reputation systems, I think these are likely to be of considerable interest to educational technology audiences.

The socio-cognitive theories of users these books draw on to explain users and community—specifically, focusing on when and how books draw on social and psychological theory to explain design decisions—for example, when a book invokes a behaviorist psychology, or draws on Maslow's hierarchy of needs, or ideas about why users should find participation meaningful. Many of these issues have emerged in my focus on the first three components, but specifically focusing on how these books are mobilizing existing psychological and sociological theory underscores what kinds of formal theoretical perspectives they are drawing from. In a sense, this focus works in the opposite direction from the other three. In the previous cases I am looking for the implied theories of motivation and social interaction in particular tactics for visual design, banning and moderating users, and designing and using reputation systems. In this case, I am looking to identify occasions when explicit social and psychological theories are invoked, and to identify the tactics or practices that those theories are used to support.

Analytic Process

Working with 28 books, most of which are over 200 pages long, presents an analytic challenge. Beyond the number of pages, my intention to retain the range of perspectives of the individual authors, along with the fact that my research questions require a close-reading approach, make the scale of this project even more formidable. Given the sheer volume of text to work with, as well as my analytic interests, in order to conduct my research I developed an iterative process of close reading and integrative writing combined with the use of a structured data sheet to collect information and direct quotes from each book. In this section, I briefly describe this process and how and why I took each step that I did. Given the importance of transparency in research process, I have tried to report each of the decisions along the way in the iterative development of my analysis.

Preliminary Analysis and Framework Development. In developing the proposal for the study, I had already engaged in extensive reading and analysis of three books: Derek Powazek's *Designing for Community* (2002), Patrick O'Keefe's *Managing Online Forums* (2008), and Anna Buss and Nancy Strauss's *Online Communities Handbook* (2009). As I prepared the research proposal, I obtained copies of each of the 28 books. I skimmed each book to confirm its relevance to the project and to inform the development of my interpretive framework. Through this process, I already had an idea of the topics and issues in the books and was confident that they were rich enough and diverse enough, but still related enough that I could use them as the basis for a study.

Initial focused analysis of highly divergent books. From the initial framework in my proposal, I began to read each of the books for the study and to transcribe sections of the books that were particularly pertinent to the research questions in my research proposal. (See the appendix for an example of the data analysis template.)

I started with books that were as different as possible within the 28 books. These included an early BBS-focused book, Charles Bowen and David Peyton's 1988 *The Complete Electronic Bulletin Board Starter Kit*; Karla Shelton and Todd McNeeley's 1997 *Virtual Communities Companion: Your Passport to the Bold New Frontier of Cyberspace*, a very technical book about a particular platform; David Mytton's 2005 book *Invision Power Board 2: A User Guide*; and a relatively recent book focused on community from an open source perspective, Jono Bacon's 2009 *The Art of Community*. With this initial analysis in hand, it was clear that there was already a considerable amount of rich data to work with—so much so that I became concerned about getting lost in the data if I just kept filling out these data sheets for each book without trying to work more systematically to begin to integrate my research.

Iterative theory/framework development. Given the quantity of this information and the need to retain a distinct perspective on individual authors, I decided early in the process to start integrating my reading of the books, and I began putting together a framework for writing up the research results. This enabled me to begin to systematically build a perspective and interpretation that I could continually refine in dialogue with the remaining books I needed to analyze. I could read the remaining books against my developing perspective to refine how I was interpreting them.

Working in this iterative approach pushed me to begin thinking about how my analysis would ultimately be assembled. From my proposal, I was unsure if it would be better to work chronologically in my writing and analysis than thematically. From reading the first four books, I could already see the advantages of each approach. On one level, it was difficult to even talk coherently about BBS tools and the assumptions operating in discussions of them with books delving into the features of reputation systems for the web. With that said, my primary interest was to uncover exactly those kinds of similarities and differences across seemingly disparate systems. From my close reading of three books in the proposal process and four in the beginning stage of my full research process, I could see the value of these two approaches.

Reflecting on the issue, I decided that there were particular values that the historical approach could serve in further framing and focusing a separate thematic approach. In short, I decided to do both. A focused historical analysis could inform the development of a more thematic treatment. Specifically, definitions of what online community is had clearly developed and shifted over the course of time in the books, and I could document and describe that. With that in hand, I could then use that analysis to inform a more systematic approach to particular functionality in these different systems. I decided to focus on answering my first research question, to identify what community meant, and what psychosocial theories of community were evident in the books in the historical account. In this respect, I could highlight how the notion of online community developed and changed over time. The second research question, which focuses on the particular tactics and features that are used to enact that idea of online community, is best addressed thematically, exploring different kinds of functions that I had already identified in the proposal.

At that point, I sketched out a rough framework for four historical periods I thought the books were falling into, and for four primary areas I thought I would concentrate on analyzing thematically. These each became the basis for the sections of my analysis, with the exception of one thematic area. I had thought it would be valuable to write a thematic section on creating explicit policy and rules posted on online communities; however, in practice it wasn't featured in enough of the books to be substantive enough for me to fully develop.

Next set of focused book analyses. Based on the categories I had selected, I chose the next seven books that I felt would best round out initial writing on each of the areas of focus. I read and filled out analysis sheets for

Mark Chambers's *Running a Perfect BBS* (1994), Cliff Figallo's *Hosting Web Communities* (1998), Amy Jo Kim's *Community Building on the Web* (2000), Stoyan Stefanov, Jeremy Rogers, and Mike Lothar's *Building Online Communities with phpBB 2* (2005), Christian Crumlish and Erin Malone's *Designing Social Interfaces* (2009), and Tharon Howard's *Design to Thrive* (2010). In selecting these next books for analysis, I tried to pick those that represented the diversity of the time frame I was working with, as well as the diversity and level of technical detail represented in the books. Having taken notes on each of these with a framework in mind for writing up my analysis, I was then able to use my notes on each book to begin drafting the sections.

Initial drafts of each section. Writing and drafting interpretations of these works has been a continuous part of the research process itself. With the outline of my ideas in hand, I worked to synthesize and juxtapose the analyses I had done into each of the initial sections. Inevitably, this involved my also having each of the books at hand as I worked to reconcile and document emergent differences between books dealing with particular points in the historical narrative. At this point, I had engaged with and taken notes on 14—or half—of the total set of books. Having notes and extensive quotations from each book at hand made it relatively easy to make statements and claims grounded in one book and to compare the directly related and relevant books to either confirm or complicate interpretive claims.

I worked through the historical sections first because the historical analysis could help to frame, focus, and set up the thematic analysis. As a result, I drafted an initial set of conclusions based on the historical analysis sections, which then informed my writing on the functional/thematic sections. A revised version of that write-up persists as a set of remarks at the conclusion of the historical section (chapter 5) of this book.

Reading sections against the remaining books. Once I had developed these initial analyses, I identified the remaining books that were likely to be particularly relevant to each of the sections. For example, the remaining two BBS books were identified as relevant to the section on BBSes. As a result, I read them to specifically look for points or issues that would counter or validate any of the points and themes I was making about them in my analysis, and as I found those points that were relevant or countered points in my draft section, I directly integrated discussion of the book into the draft. I had originally intended this to be a way of ensuring that I was taking an integrative approach and developing my ideas. It was a tactic for reducing the data I was working with into chunks that were meaningful and retained their coherence from their original authorial perspective. In practice, I found this approach to be a useful method for refining my ideas, as well as serving as a validity strategy. I was effectively testing the theories I had developed from one set of texts against a related but distinct set of texts. I used this strategy to work through the remaining sections, the result of which was a full rough draft of my analysis.

Composing the discussion section and conclusions. Once the entire rough draft of my analysis sections was complete, I assembled them into a single document. This required me to write up introductory material for the sections and to revise and refine them to ensure that they were still focused on the central questions of the study. Assembling these analyses and revisiting the initial literature review of the research proposal provided me with an opportunity to reflect and to draft the discussion section and conclusions.

Consideration of Validity Threats

While validity is a property of inferences from evidence and not a general property of research design (Maxwell, 2011), I anticipated and developed strategies and approaches to counter anticipated threats to the validity of my interpretations and arguments. To this end, I have presented my response to a series of challenges or threats I could imagine being levied against my research design upfront. These—as well as my responses to them—explain the inherent limitations and explanatory power of the analysis and results. For each, the italicized text states the threat, and the following text responds to it.

Relationship to Actual Online Communities

What does anything in these books have to do with what is happening or has happened in particular online communities? The approach to this topic loses some of the direct value and ecological validity that one would get from studying a particular community or a set of individual communities. With that said, and given my interest in the discourse that creators of these communities are engaged in, these books provide direct access to that discourse. Since I am interested in the network of ideas and theories about how to design and create these communities, I am focusing on places where these ideas are made explicit and published. Books for people who want to create these kinds of communities are a great source for that kind of information.

The Stories in the Books Are Rehearsed

The guidebooks do not represent what people who implement these tools think; they are at least one level removed from the actual thoughts of implementers. Further, the kind of post hoc ideas about the software that come from the guidebooks do not represent what the authors thought at the time. They have been extensively thought through and processed into the stories they present and thus do not accurately represent what they thought at the time. The guidebooks present widely broadcast sets of ideas about online communities. As such they do not represent any kind of normal or average implementer's ideas. Nonetheless, these books have generally been written by individuals who have successfully built and managed communities. In this respect, the anecdotes they share are valuable sets of information about their experiences. Further, recognizing these books as the work of boosters, this material is in some ways more important than the thoughts of any individual community implementer, since it is the advice that anyone looking to start a community site would find.

That point is important. The books do not offer direct insight into implementers' beliefs and ideas. The anecdotes presented in the guidebooks cannot be thought of as authentic, firsthand accounts from normal developers and administrators of online communities. Instead, they must be understood as illustrative examples. There will be legitimate kernels of the original experiences of these individuals in these stories, but that is not the real value of the stories. These stories themselves are important cultural tools for defining how these systems should work and what roles each individual plays in those stories. To that end, it is important to treat the stories less as accurate reports of historical events and more as mobilized experience reified and distilled into cultural scripts that accomplish work. Since the ideas in these works were intended to be *used*, they are more likely to represent the authors' actual beliefs.

Authors of Books Aren't Normal Designers/Developers

People who write these kinds of technical books are not representative of "everyday" designers and managers of online communities, so wouldn't it make more sense to interview more "everyday folks"? I am not primarily interested in these books as accounts of designers' thinking, as previously mentioned. Rather,

I am interested in these books as key parts in the discourse of design and management of online communities. While these authors are not average folks working on these kinds of communities, they are important voices in the field, and their books are the references that everyday folks would choose to consult. With that said, this is an important limitation. Most people who set up online communities do not write books about setting them up, so the perspectives in these books should not be construed as being universally shared. The ideas in the books do, however, come from individuals working in this area and are also being broadcast to others in the field through these publications.

Representing Distinct Authors' Perspectives

Twenty-eight books is a lot of different authors' voices. How can you be sure you are adequately representing each author's individual perspective? Each of these books represents an individual author's take and perspective, and doing justice to 28 different voices over 22 years is a challenge. My analytic process involved two explicit components designed to help try to address this issue. First, as I started to develop a sense of thematic points and apply them to the books, I would go back to each book to seek out discrepant evidence. Before categorizing a book as representing one perspective or another, I worked to identify what kinds of statements and claims in the books would invalidate the claim I wanted to make, and I searched the sections where one might find that kind of information. This approach helped me to qualify and refine claims about the texts. In addition, when I put together my analysis, I worked to recontextualize my presentation of an author's work in the context of his or her given project and goals. Most of my reporting involves substantial retelling of authors' ideas, using extended quotations to present the rich data for readers to interpret themselves.

Respecting the Diversity of Divergent Perspectives

How can you be sure you are representing the diversity of perspectives in the books? Given that I have purposely selected these books to represent different moments in time, different professional perspectives, and somewhat distinct but related audiences, I have worked to make sure that that diversity of perspective is represented in my analysis as well. So, to make sure that diversity is present, I have worked to include some analysis and discussion of each of the books or at least those representing a diversity of distinct perspectives on each topic, and to explicitly look for divergence in perspectives within the texts I have analyzed.

Why Should I Trust Your Interpretations of These Authors' Ideas?

This kind of qualitative research requires a significant amount of trust in the researcher. Why do you warrant this trust and what are you doing to manage your own biases? As previously mentioned, I worked as a community manager for the Zotero open source software project's online community for years. Given this experience, I approach these books as someone who comes from their intended audience, and, inevitably, I interpret these books through that experience. I see this as an added value to the project. As someone in the target audience for these books, my own reactions and thoughts about the books are a valuable part of my ability to interpret the works.

In practice, one of the biggest challenges in analyzing these how-to books is that they make significant assumptions about the reader's technical knowledge, so much so that even their "clarifying" analogies can be difficult to parse. For example, one of the books uses the notion of "read" and "write" in hard drives as a way of explaining different kinds of online communities, and several of the books take ideas and language from computer networking and relational databases to describe relationships between members of online communities. In both cases, without insider knowledge about these idioms, it would be easy to misread where these ideas originate. In short, my experience as part of the audience for these books helps me to take on the role of a translator/interpreter.

My experience is also something that raises some concerns. As a practitioner in the field and as someone who has researched online communities as informal learning sites, I am clearly no longer an average member of this community. To this end, it is particularly important for me to clearly support any claims and arguments I make with evidence from the texts. Further, by explicitly seeking out counter examples and negative cases, I have worked to ensure that I am not simply cherry-picking cases to support a particular argument. Again, my approach to trying to represent and share as much context and perspective from the authors of the books is intended to invite a reader of my research to get relatively direct access to the texts. To be sure, the organization and structure of this work is directly informed by my perspective and approach, but similarly, I have tried to make the process of developing that structure as transparent as possible.

Writing Up Results and Analysis

In the iterative process of reading and analyzing the books for the study that this book is based on, I decided to take two parallel approaches for presenting the results of this research. My analysis is primarily divided into two sections, one chronological (chapter 5) and one thematic (chapter 6), according to different kinds of functionality in the software supporting on-line communities.

Chapter 5, "Rhetorics of Online Community," is a historical presenta-tion of how the idea of online community developed over a 40-year period, from the earliest BBSes to the present. The chronological approach allows me to establish and illustrate tensions between different notions of online community as they developed over time. The result of this presentation is an argument for understanding the logic of online community not in terms of governance and social contracts but in terms of permission and control. This section is primarily concerned with establishing the ideology and values at play in the definitions of online community (Research Question 1).

Chapter 6, "Enacting Control, Granting Permissions," examines how the owners and administrators of online communities get their users to do what they want them to do. This section specifically explores and documents the role that visual design/information architecture, moderation tools, and repu-tation systems play in enacting control. This section is primarily concerned with establishing the tactics by which the ideology and values of online com-munity are enacted (Research Question 2).

Note on Technical Jargon and Vocabulary

As technical books are generally written for an audience of information technologists, the amount of assumed shared knowledge and context in these works is significant. Throughout both the historical and thematic sections of my analysis, I have done my best to make that technical language and discussion of systems clear to a reader who lacks insider knowledge of these systems. With that noted, I have also attempted to present these authors' perspectives as authentically as possible, retaining their words and explicating them. The results of this approach are inevitably somewhat uneven, but I have worked to focus in particular on explaining and contextualizing concepts and terms I see as most critical to the issues at hand in this context. As a result, I would ask readers to try not to get hung up on particular terms and concepts that may be less fully defined than others, as they are likely not particular points of analytic focus.

COMMUNITY AND VALUES:
A WORKED EXAMPLE OF ANALYSIS

Working through thousands of pages of text from 28 distinct books spanning decades requires that my analysis and interpretation be presented concisely. In the next two chapters, I present snippets from the books about online community and interpret them in context. However, throughout the following chapters I do not go into the depth of analysis that some accustomed to discourse analysis may be familiar with. In the interest of being concise, throughout the analysis I present my interpretations and avoid detailing the full series of interpretive maneuvers that have been involved in the analysis.

Before getting to that analysis, I have provided an in-depth exploration of one paragraph from the 2005 book *Invision Power Board 2: A User Guide*, by David Mytton. My reason for going in-depth on this is threefold: first, to provide documentation of my analytic process and thus offer additional transparency; second, to offer a model for others to follow in applying James Paul Gee's (2005) discourse analysis techniques to studying technical software writing; and third, to provide a detailed example to open up many of the key themes and issues that follow in my analysis across all the books.

For the most part, I have made use of a subset of the 27 analytic techniques Gee describes in *How to Do Discourse Analysis: A Toolkit* (2010). Below I will describe how one can apply some of those tools to a paragraph from page 6 of

Mytton's *Invision Power Board 2: A User Guide*. I have broken the statements into lines, a tactic Gee describes as his Tool 12, in order to draw attention to particular parts of the explanation and to enable further exegesis. On page 6, Mytton briefly explains the purpose of online communities as follows:

1. A community adds extra value to almost any website.
2. One of the main goals of website **owners**
3. is to keep **visitors** returning
4. for more **content**.
5. Launching a bulletin board (also known as a forum)
6. can have that effect
7. —**members** participate in discussion about
8. (but not limited to) the topics *they* are visiting the website for.
9. This can provide extra help,
10. answer questions and
11. introduce another support channel
12. to *your* website.
13. All of this adds to the "stickiness" of *your* website.

Functionally, all of the tools in Gee's approach to discourse analysis are questions to ask of a text. Thus I will phrase these as questions and work through the answers. To begin with, how is the author of this text positioned, and what is assumed about his relationship to the reader? For this question, Gee's first tool, the Deixis tool, is useful. Deictic expressions, pronouns, and terms about spatial relationships situate and relate an author to the reader and to others. In this case, the italicized terms in lines 12 and 13 establish who we, the readers, are assumed to be. In referring to "*your* website," the author locates the reader on one side of a line drawn between us and the *they* identified in line 8. Mytton, the author, is speaking to us, the readers, as people like him who run websites for *them*, identified as people visiting our websites. Significantly, this sets up a basic *us-and-them* dynamic visible throughout the texts. Even at this level of explanation, we see a dichotomy forming between two kinds of people in these texts.

From this *us-and-them* dynamic, we can step back to look at the subjects here (Tool 4). The subjects tool asks us to consider what the subjects of a text are and what is said about them. The text describes a few kinds of people— owners (line 2), visitors (line 3), and members (line 7). Conceptually, the idea of members of a community makes a lot of sense, since communities are made up of members. Similarly, the idea of there being visitors to a community makes sense, as someone might visit a community but not have established the commitment and participation to become a member. With that said, the *us* of this text, the people that the author identifies with as the writer and identifies us with as the readers, is as owners.

The idea of "owning" a community is much more at odds with the everyday notion of community. Tool 9, "Why this and not that?" is useful here. Mytton could have used words other than "owner" that have a more natural fit with conceptions of community. He could have talked about leaders or elders, but he doesn't. These terms do get used from time to time in some of the rest of the online community books, but they are terms for particular kinds of members—that is, kinds of *them*, not kinds of *us*. So if the term "owners" is consistently used instead of "leaders" or "elders", terms that fit much better with our conceptions of community, why would that be the case?

Owners and members are thus constructed as identities in the text. They are roles that the reader will play and that the people who participate in the reader's site will play. The identities tool (Tool 16) prompts consideration of how these identities are described and recognized, what it is that an owner wants, and what it is that a member wants. A related tool, the relationship tool (Tool 17), is useful here as well. Lines 3 and 4 establish the relationship between the things the visitor wants. The visitor comes to your site for content. When members in your discussion board answer questions (line 10) or offer additional support (line 11), they are providing more of this content. As the site owner, the act of running a community on your site gives you the ability to get more of the content that users come to your site for without paying for it.

The text also explains what it is that we the readers, as owners, want. Line 2 establishes that one of the main goals of a website owner is to keep visitors returning. This first line is directly connected to line 13, which closes the paragraph. All of the things mentioned in the paragraph build "stickiness," a strange term to find in a text like this for those who haven't been reading a lot of books about running online community websites. Picking apart particular pieces of vocabulary, seemingly out-of-place words, is Gee's Tool 8. In the vocabulary of marketing on the web, a sticky website is one that you keep coming back to; linguistically, stickiness is a property of the content. Content sticks to visitors. Content is the actor or the agent in this terminology, and the visitor is acted upon. While one could imagine thinking about how a site empowers people to connect with one another, or share their ideas, the objective of stickiness eschews consideration of the visitor as someone with agency and innate value and instead operationalizes him or her as something to try to get content to stick to.

So, as the owner of the community, you want to stick content to users to keep getting them back. But why do you want them to visit more? Line 2 established that it is one of the owners' "main goals" to get visitors to keep returning to the site, but why? Line 1 establishes that communities "add value," and if you ask what value, the rest of the text underscores that this is about additional site traffic. In short, the value provided is in the numbers of visits. This requires some outside knowledge from the text. All texts exist in figured worlds (Tool 26), that is, sets of assumed shared values and ideas that are presumed to be consistent with those of the reader. In this case, stickiness is valuable because it results in more visitors and web traffic, and as someone who owns a site, you want to have bigger and better numbers to boast about. Those numbers either turn into dollars directly through services like Google adwords, where visitors' clicking through ads on the site results in small direct payments to the site owner, or through numbers and figures that can be used as ways to justify the financial value of the site you own. Conceptually, if there is one thing that "owners" can do, it is to sell what they own. Thus the "value" in line 1 and the "stickiness" in line 13 are about monetary value. The community is owned property, and its owner wants to do things to increase the value of that property.

All of this starts to sound very different from our commonsense, shared notion of community. The situated meaning tool (Tool 23) brings this issue to the foreground. Readers attribute different meanings to the same words in different contexts, so in this context community has come to be something substantially different than its everyday usage outside the web. Consider the statement in line 1: "A community adds extra value to almost any website." In this sentence, community is functionally a thing that you bolt on this other thing called a website. So you own the website and you own the community. In everyday usage, communities are social relationships that emerge between people in a context, generally resulting in governance structures that maintain and sustain interaction around both shared and diverse beliefs and practices. The situated meaning of community in these texts is fundamentally different than our broader cultural notions of the term.

Focus, then, on the term "forum," which is made synonymous with online community in line 5. For many, the use of the term forum for online communities is so common, and as a description of a physical space so uncommon, that it's easy to forget what applying that term to a virtual space must have originally meant. If we try the "make them strange" tool (Tool 3) and think about the term's historical meaning, we can see how a vision for community on the web has drifted away from some of the original word borrowing that occurred. A forum is a public place for open discussion; it is functionally a piece of infrastructure in which discussion and discourse take place. While someone might own an actual physical place called a forum, he or she is not the owner of the discourse that occurs in it. What is strange about our discourse and interaction becoming computer mediated is that as it is mediated—as text is entered and saved on particular servers—the ownership of the infrastructure through which the interaction occurs simultaneously becomes ownership of the actual statements that occurred.

Using some of Gee's tools for discourse analysis, I have raked over these few sentences to explicate a considerable quantity of underlying assumptions and ideology, both stated and implied, in a short paragraph. From this example, it is already clear that in these books, which are about creating online communities, it is important to check many of our assumptions about what exactly community means. In the next chapter I will work chronologically through these books to illustrate how ideas about community have developed and changed over time.

· 5 ·

RHETORICS OF ONLINE COMMUNITY: A BRIEF HISTORY

Community Memory, arguably the first computer bulletin board system, was launched in 1973 in Berkeley, California. Its brochure explained that "strong, free, non-hierarchical channels of communication—whether by computer and modem, pen and ink, telephone, or face-to-face—are the front line of reclaiming and revitalizing our communities." In the intervening 40 years, considerable ink has been spilled over how community can or cannot exist online. The visions, rhetoric, and descriptions of online community over this period allow one to step outside the current moment and unpack the ideology behind how we currently think about online community.

The term "community" brings to mind notions of belonging, shared interests and values, and governance. The goal of this chapter is to provide a concise history of some of the visions and tensions in how online community is defined and developed in these technical books. Technical books and how-to guides for starting and managing online communities offer a window onto how ideas of community have been negotiated and defined in this 40-year period.

In providing this history, I focus on how definitions and visions have developed in a series of distinct periods: the BBS era, the early web, the emergence of online community platforms, and a recent shift to using the term "social" instead of the term "community." Each of these periods has witnessed significant changes in how community has been defined and developed. However, this chronological story does more to clarify tensions between a set of persistent themes evident to differing degrees in each of these periods.

In the books analyzed, online communities fall largely into a spectrum of definitions between two visions. At one end, online communities are envisioned as utopian places where frictionless forms of "electronic democracy" can flourish. Taking the notion of "electronic democracy" seriously, users in these online communities would be citizens. The other set of texts presents community as something that is "owned," where members are not so much members, or even customers, but are instead reduced to "users." In many cases, these opposite ideas of online community can be found in different parts of the same book.

Over time, a technical definition of community has emerged and flourished. "Community" is often expressed as a set of features in a particular application, or something that is created by a set of technical features. For instance, when one author explains that the phpBB web forum software "allows website owners to add a community to their existing site within minutes" (Mytton, 2005, p. 1), community is a feature set, not something that forms over time through social interaction. In either case, over the course of the history of these books "community" itself increasingly becomes the object of design. It's not that the goal is to design functionality of websites; rather it is to design a set of features that prompts people to interact as the designer would like them to.

Below I provide a graphic organizer of the set of shifts and developments over time in discussing online community in these books. It's an oversimplification, but it can serve as a point of reference to help keep a reader from getting completely lost in the details of the particular historical sections. From left to right, the diagram presents a spectrum between the democratic/communitarian impulse to think of online community as something that emerges through social relationships and the development of norms, rules, and governance of members, and the controlling/authoritarian/corporate impulse to think of online communities entirely as the property of the site creator/designer/administrator. From top to bottom, I present the following sections based on the eras I have identified in the history of these books.

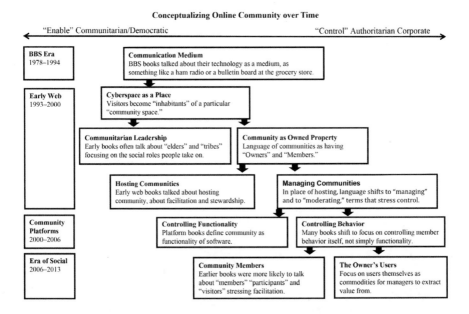

Figure 5.1. Conceptualizing online community over time. A diagram illustrating themes identified in this book and how they map onto communitarian and authoritarian polarities over time.

The BBS and the Sysop (system operator). "Finally, into our hurried, harried lives now comes a computer solution that is compelling because its simple concept is already familiar to us: a computerized bulletin board." So opens Charles Bowen and David Peyton's 1988 book *The Complete Electronic Bulletin Board Starter Kit*. From the late 1970s through the mid-1990s, computer bulletin board systems were the platforms on which the features of online community software were developed and defined.

There aren't many published books from the early phases of BBS development. Tom Mack, one of the creators of the RBBS-PC (Remote Bulletin Board System for the Personal Computer) software application, explains in the introduction to *The Complete Electronic Bulletin Board Starter Kit* how grateful he is to have this book out there explaining how to use the software he helped create. He is thrilled that "users no longer have to depend on contacting" him on the phone to answer their questions about the RBBS-PC. As he explains, "the biggest fan of *The Complete Electronic Bulletin Board Starter Kit* will be my wife and long-suffering supporter, Sheila, whose license plates read 'PC-Wido'" (Bowen & Peyton, 1988, p. x). While BBSes had been around for a decade, they weren't written about for general publishers until this 1988 book. The publication of *The Complete Electronic Bulletin Board Starter Kit* illustrates the transition of the BBS from a fringe technology to a mainstream one.

The conventions that define online communities and the posting and sharing messages in threaded discussions were established in the era of bulletin board systems. Electronic bulletin board systems had a practical utility but also came to represent larger ideologies of an information future. These systems consisted of software installed on an individual computer that other users could connect to through a modem. The individuals who ran these systems were called system operators, or sysops for short. Whereas one now talks about online communities having users, the users of BBSes, given the nature of their connection to the discussion board, were referred to as callers.

Many of the books explaining how to set up and run online communities on the web in the last 20 years begin by explaining the roots of their systems in BBS software. As such, it's important to contextualize any discussion of online communities in consideration of BBSes. BBSes were first and foremost places where callers could read and post text-based messages.

From the caller's view. For many readers whose experience of the web is limited to recent years, it will likely be difficult to imagine exactly what using these bulletin board sites was like. The following description from Mark Chambers's 1994 book *Running a Perfect BBS* can help provide a sense of what connecting to different BBSes would feel like.

> Think of a typical evening spent calling other bulletin boards; your first call during an online evening might be to a local gun enthusiast's hobby system, where you might join in a discussion for or against gun owners' rights. The next call might connect you to a long-distance profit system; if you subscribe, you probably spend the duration of the call listing the new files made available during the last week and downloading the files that catch your eye. Finally, you might login to another local board patterned after a coffee house that specializes in online chat and FidoNet messages. You'd probably find five or six users conversing with the sysop in the Chat Lounge, and you might join in before you downloaded your FidoNet mail using the system's QWK door. (Chambers, 1994 p. 19)

Each of these individual BBSes would likely be running different pieces of software, configured and customized toward particular ends. Aside from making separate calls to each of these distinct BBSes, the story above reads very similarly to how someone might now use the web: going from site to site to engage in a discussion of gun rights, downloading files, and checking email—the important difference being that here a user would call into these discussion boards one at a time, while someone today could well have a range of different sites or forums open in different web browser windows or tabs.

BBSes as part of an online world, not distinct online communities. Books about BBSes do not use the term "online community" to describe these systems. The idea of these systems as online communities is a label that has been retroactively applied to them. Chambers (1994) talks about an "online world," but in doing so he describes the entirety of the world of different BBSes he might call up from his home. The term "online community" did not broadly come into use until the late 1990s.

BBSes as a communication medium. Instead of describing BBSes as platforms for community or as communities themselves, BBSes were largely presented as mediums for communication. In *The Complete Electronic Bulletin Board Starter Kit*, Charles Bowen and David Peyton provided examples of the range of practical uses a BBS could have. For example, "A hospital sets up an RBBS-PC as a community service," or "A group of local realtors" use a BBS to share info on a private board, or "A mother and father with college-age children...create a BBS in the basement as a never-closing communication line for the kids," or "A public school system creates a multiuser statewide network to discuss education issues with parents, teachers, and students" (Bowen & Peyton, 1988, p. 5). In practice, these bulletin board systems provided valuable and practical opportunities for quickly sharing information. In the same vein, John Hedtke, in his 1990 book *Using Computer Bulletin Boards*, describes the BBS by way of analogy to other communication media. He explains: "BBSes are used in many different ways." For example, "Like a cork bulletin board, they can be used to post pieces of news, humor, and notices." However, "Many BBSes are like CB or ham radio, you can have a conversation with one or more persons by typing messages back and forth to each other" (Bowen & Peyton, 1988, p. 2). In these explanations, BBSes are a communication medium, not platforms for community, or defined as community in their own right.

Understanding BBS nation as the online world. With that noted, in the 1992 book *Bulletin Board Systems for Business*, Lamont Wood and Dana Blankenhorn did go on to define the collective network of BBSes as part of "BBS Nation." In their words, "In opposition to services such as CompuServe, Prodigy and BIX" there are "millions of newly minted computer users" who have "preferred to dial into BBSes set up by their neighbors, leading the growth of a (virtually underground) BBS nation" (Wood & Blankenhorn, 1992, p. 5). While there were a range of systems such as Prodigy and

CompuServe establishing networks for individuals to communicate, discuss, and share information, Wood and Blankenhorn were thrilled to note that in 1992 the underground network of BBSes served as a more grassroots nexus of communication. In this sense, the BBS nation functioned as one large online community, a network of networks of discussion and information.

Individual BBSes were connected together through systems called Echo Nets. As David Wolfe explained in the 1994 book *The BBS Construction Kit: All the Software and Expert Advice You Need to Start Your Own BBS Today!* "Even a small BBS can be hooked into a large echo net (a network designed for the sharing of messages between a vast number of BBSes), which will give users the ability to talk with people around the world" (Wolfe, 1994, p. 1). In this sense, BBS Nation already looked a lot like the World Wide Web, which at that point was coming on the scene.

The peak of the BBS coincided with the emergence of the technology that would mark its end. Alan Bryant, in the 1995 book *Growing and Maintaining a Successful BBS: The Sysops Handbook,* notes that "it wasn't until the 1990s that BBSes really took off," when "Recognition in the mass media, expanding modem speeds, and continuing improvements in software like graphical interfaces, caused the BBS industry to light up," often "riding on the coat tails of the Internet" (Bryant, 1995, p. 3).

In considering the future of the BBS in 1995, Bryant suggested that it was important to think about "what they accomplish" instead of "how they accomplish it." He identified the features of the BBS primarily as platforms for "messaging, transferring files, displaying information, and gathering information." Based on these features, he explained that many "commercial online services (such as CompuServe and America Online) as well as some Internet service providers could be called bulletin board systems if you chose to look at them that way" (Bryant, 1995, p. 4). In this sense, the functionality of the BBS lives on in the various platforms that have come to replace it in usage. In Bryant's words, "some people think the web makes a better BBS than a BBS does, especially when combined with other Internet features" (Bryant, 1995, p. 5). In considering the future of the BBS, he asked rhetorically, "Will there be better connectivity between BBSes and the Internet than exists today? Definitely. Will BBSes go away? I doubt it. No matter what the place or nature of the changes that are coming, bulletin boards will remain an important part of the computer communications scene for decades to come" (Bryant, 1995, p. 14). If we take the broader definition of the features, then Bryant's prediction has come to fruition. There are indeed many online discussion and bulletin board systems running on the web today. With that said, the bulletin board that individual users call up rapidly became a thing of the past in the latter half of the 1990s.

BBSes as electronic democracy. For the most part, books about BBSes describe them as a practical tool for sharing messages. In 1988, Bowen and Peyton explained that the BBS is a "practical solution to assorted 'real-world' communications problems" (Bowen & Peyton, 1988, p. 5). However, the authors also make much bolder claims about what BBSes represent: "bulletin boards like RBBS-PC are also an encouraging example of electronic democracy: they provide a low-cost means for every PC owner to have his or her own soapbox and an audience with the nation. Also, they illustrate how strangers can cooperate in the free exchange of information" (Bowen & Peyton, 1988, p. 5). Here democracy becomes everyone having his or her own soapbox— that is, everyone who can set up and run his or her own system.

For context, the idea that you could set up a communication network for sharing messages on your computer was powerful, but what aspect of it made for electronic democracy? One had long had the ability to call just about anyone with a phone, so what made the BBS "democratic"? Already in the era of the BBS, we see the notion of computer-based networked communication as implying the potential for a new kind of social interaction. The disjunction between the practical focus of a technical book about running a BBS to stay in touch with your family and the idea that the BBS embodied and empowered some new mode of democracy suggests that imagination was already well ahead of the state of the technology. The future potential of the technology was already focusing attention on the kinds of online communities we would create.

Given that visions of online community were already taking shape in a notion of electronic democracy, what kinds of norms and rules were emerging to ensure that this space was fair? In discussing the configuration options of a particular bulletin board software package, Bowen and Peyton make suggestions for how to configure certain options that would serve as gatekeepers, letting in particular people and keeping others out. The software came with a configuration feature that allowed the sysop to "Deny access to callers who use 300 baud." As they go on to explain, "Frankly, some sysops think that those logging on at 300 baud are more likely to be juvenile in their outlook and thus leave childish messages or try pranks. (Of course, this theory is weakening as the price of faster 1,200-baud modems comes well within the reach of us all" (Bowen & Peyton, 1988, p. 75). By setting this configuration option, the system could automatically deny access to anyone who had too slow a modem, with the stated belief that people with slower modems were younger individuals or people who aren't keeping up with newer modem technology. Bowen and Peyton "suggest you answer no to this question and allow in all new users," as "you still reserve the privilege to shut that gate later by revising this parameter, so why not give everyone a chance at first?" In short, they provisionally suggest leaving the system open. Here we see the logic of this particular system and those that came after it in terms of control and management.

As one reads through the configuration options of any particular BBS, it is clear that these kinds of configuration options enable the sysop to set up the baseline rules of the system. One can decide to set an option to let in people based on the way the system sees those who connect to it. In this case, because the modem's speed is given to the sysops system, that information becomes a way for a sysop to make judgment calls about the kinds of people that are represented by the differences evident in the signal their system receives. This is indeed a place for users to create their soapboxes, and they are free to choose who can and cannot connect to that particular soapbox.

BBSes and free information ideology. "The idea behind BBSes was a sharing of information for free," Bowen and Peyton explain early in their book (1988, p. 6). Furthermore, "The spirit of openness was imbued in telecomputing from the very beginning" and, more specifically, "RBBS-PC itself carries on that philosophy, with the free exchange of information being one of the principles behind—and an essential means to this remarkable program's development" (Bowen & Peyton, 1988, p. 7). It was still 10 years before the term "open source software" was coined, but BBS software that Peyton and Bowen shared was already a part of the free software movement taking shape in the 1980s.

Many of the basic tools of online community have their roots in the free software movement, as do many of the values that govern and organize how their designers and administrators think about information. This free software and free information undercurrent persists in thinking about what should and shouldn't happen in online communities. While this remains an important undercurrent in the development of online community, it is rarely stated in the same way it was in Bowen and Peyton's 1988 book. From their perspective, the spirit of openness was a key part of telecomputing from the beginning. This vision of information and communication informed the development of norms and values in how these technologies have been used.

Community on the web takes off. The late 1990s saw the publication of a series of books explaining how you, too, could create your own online community. In the 1997 book *Virtual Communities Companion: Your Passport to the Bold New Frontier of Cyberspace*, Karla Shelton and Todd McNeeley explained that "the Internet is a powerful medium for much greater things than just archiving data; it's a place to interact, to live, to build community" (Shelton & McNeeley, 1997, p. xxiv). Similarly, in the 1997 book *How to Program a Virtual Community: Attract New Web Visitors and Get Them to Stay!* Michael Powers explained: "A virtual community is simply an electronic meeting place where a group of people gather to exchange ideas on a regular basis" (Powers, 1997, p. 3). Where the BBS books described an "online world," or other previous books might have described "community networks" that connected members of existing communities, the idea of "virtual community" and "online communities" created as a result of the networked technology itself is a primary feature of these discussions of the early web.

The idea of online community was anchored in the notion of "cyberspace" the now out-of-fashion idea of a virtual cyber environment that functioned much like physical space. Discussions around cyberspace took on a kind of spiritual rhetoric, and the idea of online community was imbued with high-minded rhetoric. Coined by William Gibson in 1982, and made popular in his 1984 science fiction novel *Neuromancer*, cyberspace offered a vision of the future of networked human interaction. With that noted, the concept of cyberspace does not show up in any of the BBS books. The books about online community from the late 1990s are the ones that operationalize this notion from science fiction and use it to define online community. In this operationalization of cyberspace, the inception of online community had a hierarchical and possessive mentality, there were "owners" and there were "members," and the somewhat utopian vision of finding meaning and belonging in these online communities coexisted with the idea that these virtual hubs of belonging would be owned and managed for the benefit of whatever company decided to establish them.

The language of networking in the language of community. In a chapter of the 1997 book *How to Program a Virtual Community* titled "Meeting Your Neighbors in Cyberspace," author Michael Powers provides a general explanation of how the web becomes a place for community to happen. In doing so, he maps the language of the networking principles that enable the creation of the web onto the language of describing online community.

While an HTML Web site delivers information to visitors (forming a single Web host-to-visitor link), a virtual community provides a location for visitors to enter the Web site and form a community (creating a many-to-many link among the visitors). The visitors then become inhabitants of this community space. In this mode, the community forms around the information or material that interests them. The Web site host, by providing an interesting location, gains a community of likeminded customers, students, or business managers who regularly travel to the site to gain information and meet others. (Powers, 1997, p. 4)

We don't generally think of community in terms of many-to-many or one-to-many: these are terms from the principles of networking and databases. It is the language of a data model. As the idea of online and virtual community was being invented, it was being infused with the language, rhetoric, and vision of the networking technologies that enable it. Powers is suggesting here that the individual hosting the website has the opportunity to turn that site into something more than a "brochure"—a term he mentions elsewhere in this context—and make it into a "community space." The term "space" here is significant, as it signifies a particular vision of the web.

The chapter title, focusing on cyberspace, is part of a reimagining of the nature of networked communications. Where the BBS books talked about their technology as a medium, as something like a ham radio or a bulletin board at the grocery store, the language found in Powers's book, focusing on enabling the many-to-many relationship between visitors to your website, turns those visitors into "inhabitants" of that particular "community space." Powers is using the language of data relationships and databases as a framework for explaining what a community is. More than a mode of communication, this becomes a virtual place. This book is unique among the other books about running your own online community in that it traces the idea of online community back to MUDs (Multi User Dungeons/Domains or Dimensions): the real-time virtual chat environments dating back to the late 1970s that were often used and created as places for fantasy or role-playing games. The fantasy worlds and virtual places of these Multi User Dungeons provide a key part of the vocabulary for defining the web as cyberspace. Where the BBS was a communication medium, the web was taking on characteristics from these virtual fantasy worlds.

Spirituality in cyberspace. Some of the discussion of online community in the books of the late 1990s seems contemporary, but some of it already reads as particularly dated. Karla Shelton and Todd McNeeley begin their book by asking: Where is online community happening? "We have two answers.

First, it's obviously going on in cyberspace, where geography isn't important; where those in California become friends with those in Maine" (Shelton & McNeeley, 1997, p. xxiv). The authors value the erasure of geography provided by discussion boards. Communities are no longer bound by proximity; they are bound only by shared interest or affinity. The authors go on to further explain that "in cyberspace, it's happening on HotWired and iVillage, on IRC and AOL, and it's already happened on the WELL [see below]. It's even happing at Intel and Compaq. There is hardly an aspect of the internet that isn't affected by this new universal focus on community" (Shelton & McNeeley, 1997, p. xxiv). Here we see that cyberspace is bigger than the World Wide Web. HotWired and iVillage are two websites, but IRC (Internet relay chat) is a separate protocol for communicating by way of the Internet and AOL while providing access to the web, and also provided its own suite of chat rooms and services that were not part of the open web. While online community will increasingly become a term for web-based communities, in Shelton and McNeeley's book, as well as in some later books, it is a broader reference to any range of modes for communicating through a variety of telenetworking tools. The mention of the WELL is important, as this particular site has become a touchstone in all kinds of books about online community. I will return to reflections on the WELL shortly, but back to unpacking this notion of cyberspace in these texts.

Shelton and McNeeley ask another rhetorical question: Why is this happening? They explain: "Because, beyond the byte and the baud, past the silicon, further than the software code, on the other side of the screen, there are humans. And humans want to live together and talk and laugh and cry and feel" (Shelton & McNeeley, 1997, p. xxiv). Descriptions of the web in books from the late 1990s consistently focus on its emotional potential. Aside from being a place to "laugh and cry and feel," Jon Katz, a contributing editor for *Wired*, explains in the foreword to the book that the web can be "a powerful emotional, even spiritual, experience." Amy Jo Kim, in the 2000 book *Community Building on the Web: Secret Strategies for Successful Online Communities*, explains that the "net" makes it easier to "deepen relationships" and "meet like-minded souls" (Kim, 2000, p. x). Where BBS books had mostly just described how a particular piece of software could help you better connect your sales force, or your church, or serve as a message center for your family, books about online community in the late 1990s were selling something much more grand: a cyberspace that could serve as the platform to meet our most basic human emotional needs, a place to commune with like-minded souls.

The cover of the 1997 book *Virtual Communities Companion: Your Passport to the Bold New Frontier of Cyberspace*, by Karla Shelton and Todd McNeeley, illustrates this vision of cyberspace. The cover shows a set of glistening translucent avatars walking around in a three-dimensional town square. In the distance we see high-rises, but up close we see seven different virtual people, some walking about, one sitting in a park, and another at a table outside a virtual café. As Amy Jo Kim explains in her book, "The Web is becoming our collective town square—more and more, people are turning to Web communities to get their personal, social and professional needs met" (Kim, 2000, p. xi). While the specifications for the web set up rules for transferring information and establishing a communications medium, the authors of these books were invested in making clear that what had really been created was the placeless place of cyberspace; a non-geographic and boundary-less place in which we could find spiritual fulfillment for the soul.

The books have a guru-like quality to them, telling us about how we can all find enlightenment on the web. However, since they are written by developers and consultants who make their money providing their visions to companies, they are also about how this new "frontier" might turn a profit. Discussion of who owns these spaces brings an important contrast to this vision of a utopian, emotionally charged cyberspace.

The owners of community. "Successful communities evolve to keep pace with the changing needs of members and owners," as Amy Jo Kim proclaims in the first pages of the 2000 book *Community Building on the Web* (Kim, 2000, p. 3). While the web might provide an amazing set of experiences, the most fundamental categories of people in cyberspace according to Kim's book are the owners and the members. Elsewhere Kim refers to online communities as "town squares," but these are clearly not civic spaces. In these books, online communities are owned facilities in which members can do things only to the extent that the owners allow. The wording here is important. Owners are not stewards, or leaders of the community. Owners are not community elders (a term Kim uses elsewhere for a particular kind of member). Owners own the community. Where it might be normal to think about owning the facilities where a community meets, it is something quite different to think of community itself as property.

Kim goes on to explain that the owners are "the people who will be funding and/or running your community" (Kim, 2000, p. 10) and stresses that it is important to establish their goals for the community. The goals would be different in different situations. For example, is the community "a business venture? A labor of love? A PR stunt? A research experiment?" (Kim, 2000, p. 11). The idea of community in these terms—as a stunt, a research experiment, or a business venture—reframes some of the utopian values expressed throughout these books. These are at once experiments and PR stunts created by developing software that people use, but the software that structures, mediates, stores, and creates the experience of cyberspace is set up and run by and for people who own the software and hardware and pay the bills.

An example of how these relationships play out and how Kim suggests designing communities to meet the owners' and members' needs comes in the form of a case study she provides about an online community created to help the makers of L'Eggs pantyhose sell their product. It is worth considering this case study at length to establish how the competing interests of members and owners come together.

> It's especially important to be clear about your vision if you're trying to attract a particular audience. The makers of L'Eggs pantyhose discovered the importance of this principle in 1995 when they launched an ambitious, expensive Web site for the purpose of fostering brand loyalty and learning more about their market. To develop closer relationships with their customers, they included a discussion area called "The L'Eggs Community." Much to their surprise, the discussions quickly became dominated by men who enjoyed wearing pantyhose and were thrilled to discover an anonymous setting where they could trade tips and not feel so alone in their somewhat unusual habit. The company that financed the Web site, however, was less than enchanted with this turn of events. The women they were trying to attract were put off, and shied away from participating in the discussions. Since then, the L'Eggs Company has learned to market more explicitly to its target demographic. But the point remains: unless you communicate your purpose clearly, people will use your Web community in ways that you never intended. (Kim, 2000, p. 22)

In short, a pantyhose company created a website for its customers to communicate with each other, but when those customers turned out to be men and not women, the company was not happy. Even though the discussion boards were building a community and brand loyalty, the people in the community were not the desired people. The discussion board was shut off. Some might object to the notion of software on a pantyhose website where customers talk about the product as "community," which is exactly what online community often came to mean in the late 1990s. While this could have been quite useful to the men who wanted to talk anonymously about wearing pantyhose, and while it might have been meeting their needs, because the community itself is owned by the company, it's easy for them to just turn it off.

What advice would Kim offer to L'Eggs? How could they, as the owners of the community, have better controlled the outcomes of the community to meet their needs? She suggests this: "Imagine if the L'Eggs community had used a tag line like iVillage's 'Real Solutions for Women,' their site would have evolved in a very different way" (Kim, 2000, p. 22). In something as simple as how they described the tagline, Kim suggests they could have dissuaded these men from coming together to talk about pantyhose. Who knows if this would have worked or not, but the point reinforces the ideology of "community building" (the title of the book) that Kim envisions. Elsewhere she suggests that "It's up to you to figure out the restrictions that best meet the needs of your members and support the kind of community you are trying to create" (Kim, 2000, p. 71). The entire process of design is about establishing the kinds of restrictions, the visual and textual queues that establish who is and isn't welcome and how they should behave. As the owner of the community it is your possession, and in this description even the members are yours.

Bozos in the WELL. Books about online community frequently discuss "the WELL"—The Whole Earth 'Lectronic Link, created as a BBS in 1985 by Stewart Brand and Larry Brilliant and which is still in operation today. It calls itself the "the primordial ooze where the online community movement was born." Howard Rheingold's 1993 book *The Virtual Community: Homesteading on the Electronic Frontier* left a mark on many who have gone on to write about online community, and Cliff Figallo—one of the early authors of books about online community—had worked directly in the WELL.

It's worth digressing for a moment to unpack some of the significance of the creators of the *Whole Earth Catalog* playing such an instrumental role in defining some of the features and functionality of online communities. Published between 1968 and 1998, the *Whole Earth Catalog* listed products that promoted a self-sustainable way of life. Environmental historian Andrew Kirk (2007) argues that the publication played an important role in redefining the environmental movement, shifting away from an anti-technology stance to a pragmatic countercultural green that embraced the role that technology could play in redefining the relationship between people and nature. The catalog played a role in defining a quintessentially Bay Area countercultural movement focused on a mixture of environmental sustainability, an embrace of technology, and a strong sense of individualism. It is in this context—part green counterculture and part techno libertarian entrepreneur—that the functionality of online communities on the web was defined.

Many of the books on online community, even those published in the last few years, trace the origins of online community back to the WELL. Reflections on the WELL in books about running online communities often bring up a tool called "the bozo filter" and the adage that it was "tools not rules" that should shape community interaction on the web. These two concepts illustrate a cultural logic that was taking shape.

Shelton and McNeeley (1997) admire the WELL and Electric Minds, a site Howard Rheingold created after his experience with the WELL, for having a "framework for a policy for dealing with potential problems." One of the guidelines they quote explains: "We believe in the axiom 'tools, not rules,' so rather than tell you what to do, we give you the ability to do things, and let you decide how and when to do them." This mentality moves away from ideas like governance into a focus on individuals' asserting control of what they see and experience. They go on: "For example, we provide you with the 'bozo filter,' which you can use to make the words of a specific user disappear from your view without censoring them" (Shelton & McNeeley, 1997, p. 401). The heart of this concept is a strange kind of technology-enabled libertarianism: if we could just build the right tools to block out what we don't want to see, anyone could do whatever they wanted to do.

Hosting vs. managing. In contrast to Kim's ideas about owning community, Cliff Figallo used a term that he and his colleagues from the WELL weren't able to get to take off as successfully as the "bozo filter." Figallo's 1998 book *Hosting Web Communities: Building Relationships, Increasing Customer Loyalty, and Maintaining a Competitive Edge* focuses on the idea of hosting, not owning, web communities.

The idea of hosting is itself a central focus of Figallo's writing. "*Hosting* is an appropriate term for inviting users into a virtual location and treating them as guests. It's a service role with a purpose, which is to make the guests feel comfortable, appreciated, and, in some cases, empowered" (Figallo, 1998, p. xi). The concept of a host inviting in guests is fundamentally different, and the fact that the term appears in only two other books (Derek Powazek's 2002 *Designing Online Communities* briefly uses the term, largely because the book includes discussion of Figallo's book and an interview with Figallo, and Michael Powers's 1997 book, which I will discuss shortly, also uses the term), underscores this idea as a bit of an alternate reality for how online communities might have been conceptualized. From the conception of hosting, the people who participate in a site aren't members, and they aren't users; they are guests invited into your space to be made comfortable and appreciated.

In further defining the role of host, Figallo explains: "The hosting role can be that of a master of ceremonies, a meeting facilitator, or a digital custodian. There may be elements of entertainer, justice of the peace, and group therapist in the host's job description, but in any case, the host should be a liaison between users and site management" (Figallo, 1998, p. xiv). The notion of site management brings in the idea of the "owners," but each of the roles he describes—facilitator, entertainer, and so forth—all orient the nature of the exchange around the idea of service to guests.

The role of host is also tied to another notion largely absent from books about online community: the notion of governance. Governance, the idea of a process by which decisions are made by members of a group, is at the heart of the idea of community, but it is by and large not to be found in books about the concept of online community. The two exceptions to this offer an important place to think through a different conception of community, one from 1997 discussed here and another from 2009 discussed toward the end of this chapter.

The idea of a site owner playing host to guests or visitors also appears in the 1997 book *How to Program a Virtual Community*, by Michael Powers, under the heading "Meeting your Neighbors in Cyberspace." Here, Powers devotes a section to the notion of governance, suggesting the importance that someone interested in establishing an online community should place on working with members of the community to define and establish rules for governance. He notes that many online communities are effectively dictatorships but that there are democratic roles that these communities can take on as well.

What sense is there to the idea of corporately owned communities? A propos of the discussion of the difference between hosting and owning communities, Powers makes this insightful statement about for-profit endeavors to establish online communities, suggesting that "it will not be possible for a brand-name jeans manufacturer to host a democratic community with any true assurance that the needs of the company will not override the autonomy of the community" (Powers, 1997, p. 227). In this case, "This virtual community is just like a real life situation in which a profit-driven company provides space for a group of people." These corporate-owned online spaces are explicitly dictatorships run by their owners, a point that was not previously discussed in any book about running an online community.

The period at the end of the 1990s, during which the idea of online community was conceived, involved contested notions of the terms. That is, at this point, notions of governance and hosting, terms which are much more in line with the ideals of community we tend to bring to the term in everyday life, were viable frameworks for conceptualizing online community. It is telling that Powers insisted that "Every virtual community on the Internet, whether for profit or not, should provide a social contract in writing to inhabitants before they join the community" (Powers, 1997, p. 234). This makes sense from the perspective of someone envisioning a future of communities on the web in an authentic notion of individuals participating in explicit social contracts. The fact that the terms of service for websites, discussion boards, and social networks are generally tucked away, never to be read by users, underscores just how far from this vision of online community our situation has drifted.

The disconnect between Figallo's idea about hosting and the notion of owning communities in Kim's work is indicative of some general problems Figallo reports regarding how the term community is beginning to be used. In 1997 Figallo feared the term community was "in danger of being watered down into meaningless jargon" (Figallo, 1997, p. ix), further suggesting: "For me, community has always been an important, almost sacred term preserved for relationships that have more than trivial meaning" (Figallo, 1997, p. ix). Unpacking the substance of his fears provides insight into different perspectives about how the web could and should have developed as a community platform.

Figallo is critical of sites that aren't focused on "nurturing relationships between people and involving them in their site's content development" (Figallo, 1997, p. 28). For him, "relationships, contributions, and involvements" between participants in online communities "are central to our working definition of community on the Web." The focus in all of this is on bringing people together online, not on getting them to do exactly what you or the presumed owners of a given site want. He goes on to explain the implications of this definition.

> According to that definition, members of a community feel a part of it. They form relationships and bonds of trust with other members and with you, the community host. Those relationships lead to exchanges and interactions that bring value to members. It's that value that draws them back repeatedly to your site where, over time, they build shared histories of experiences and events. This reliable traffic and the members' contributions of information, ideas, and feedback are the major benefits you'll realize by fostering community on the Web. (Figallo, 1997, p. 28)

Those final words—"fostering community on the Web"—capture the essence of his perspective. While Figallo is interested in the potential value that online communities can bring to businesses, he positions that value in the facilitation of the development of shared history. He brings an abiding respect for what he had termed "sacred" elsewhere. This 1997 vision of "hosting," of "facilitating" and "empowering" individuals to participate and develop in online communities, is rarely found in the books that come after this. It would seem that his fears about the changing meaning of "online community," specifically about the term being watered down, were largely realized.

Cyberspace's 3D future never came. The death of the idea of cyberspace is likely connected to the failure of its future. Books on online communities written during the 1990s tended toward exuberance in prognosticating a future of persistent 3D online experiences. For example, the 1997 book *Virtual Communities Companion* includes a copy of the software to set up your own version of The Palace, a 3D environment for setting up online communities others can connect to over the Internet. In an interview, Mark Jeffrey, "director of commercial ventures for The Palace, Inc.," expresses a value evident throughout the books of this era. He explains that "in two to three years, all online communities will be avatar-based" (Shelton & McNeeley, 1997, p. 184). In the logic of cyberspace, it made a lot of sense to believe that the next phase of the web would involve a rapid shift toward creating 3D immersive modes of interaction.

The logic driving this belief in the future of 3D immersive online community software offers insight into the underlying vision of cyberspace that animated many of the books analyzed in the study that the present book is based on. After announcing that text-based online communities would disappear in a few years, Jeffrey explained: "The larger group of the population will not tolerate a dull text interface in the same way they would not tolerate the linear and confusing DOS and waited for a graphical user interface as the Mac or Windows. Also, text does not do much for the sense of *being there* that a graphical environment does" (Shelton & McNeeley, 1997, p. 184). It would turn out that the sense of "being there" simply must not have been as important as it was imagined to be. Somewhere during or after the dot-com era, or the film *The Matrix*, the boosters of online community stopped proclaiming that the future of online community was going to be all 3D virtual worlds. These visions of cyberspace couldn't have been more wrong. While the massively multiplayer game *World of Warcraft* has been incredibly successful as a game, games proved to be the exception. The excitement for 3D online communities outside of games faded.

While the platforms for 3D online communities didn't take off, a set of other platforms did. In the first years of the twenty-first century, a set of software platforms for running online communities became the default toolkit for creating online community websites. These platforms—phpBB, vBulletin, and Invision Power Board—looked much more like bulletin board systems than they did the 3D worlds that cyberspace evangelists were prognosticating.

Platforms define online community. Disappointed with the commercial Ultimate Bulletin Board System and the open source Phorum software, developer James Atkinson decided to create his own software package. As Stoyan Stefanov, Jeremy Rogers, and Mike Lothar explain at the beginning of their 2005 book *Building Online Communities with phpBB*, "phpBB was 'born' on July 1st, 2000, at 06:45 PM. We know the exact date and time because that is when James posted a message on an Internet forum saying that he had created a bulletin board and would like some help testing" (Stefanov, Rogers, & Lothar, 2005, p. 8). His work on this particular piece of software rapidly created one of a series of discussion board software packages that have since become synonymous in the minds of many with online community.

When you find a site, or a section of a site, that calls itself an "online community," there is still a good chance that it is running a version of phpBB, vBulletin, or Invision Power Board. These systems, each created around the year 2000, pulled together a set of features and functionality and packaged them together into sets of server-side software.

By the middle of the decade, the platforms had become popular enough that Packt published and marketed handbooks for each of them, and other technical books continue to offer advice on how to configure and implement these systems to create your own online community. A phpBB book describes the software as an "Internet community application, with outstanding discussion forums and membership management" (Stefanov et al., 2005, p. 1). These technical books explain how the ideology of online community that had been taking shape in the early web years was translated into functionality that would quickly become the default. The mixture of discussion forums and "membership management" lend insight into the cultural logic at work inside these applications.

Community as a feature set. On page 1 of the 2005 book *Invision Power Board 2: A User Guide*, author David Mytton explains that the software "allows website owners to add a community to their existing site within minutes." In this book, as in many of the other technical books about these applications, community becomes synonymous with a set of features. To "add" a community to a website in minutes requires us to stop thinking about community as something that people create through ongoing interaction, through the development of shared history, and to start thinking of community as an application or feature set. Another book about phpBB uses related language, explaining; "Forums are an easy and popular way to implement a community on your site" (Douglass, Little, & Smith, 2005, p. 219). In this case, and the case of other books about these software platforms, community has become a feature set. Where one might have used terms such as develop, facilitate, or cultivate in the late 1990s, by 2005 a community is something to be implemented. In these systems, community is defined by the way the software enables users to sign up, gives them a rank based on what they have and haven't done, and tracks and shows how many times they start discussions and what their permissions are to do things in the system. Community has become the interaction between a set of scripts and a database. It is the enacting of a set of computational procedures and records in a database.

Mytton goes on to tell us why it would be valuable to add a community to an existing website. His description illustrates how thinking of community as a set of features that one implements turns community members' activity into a strategy for generating additional site traffic.

> A community adds extra value to almost any website. One of the main goals of website owners is to keep visitors returning for more content. Launching a bulletin board (also known as a forum) can have that effect—members participate in discussion about (but not limited to) the topics they are visiting the website for. This can provide extra help, answer questions, and introduce another support channel to your website. All of this adds to the "stickiness" of your website. (Mytton 2005, p. 6)

As Mytton explains it, the site administrator wants to add a community to her website because it brings valuable web traffic. The website owners want people to frequent the site because they want people to have more chances to view the site content and advertisements. People come to the site and talk about what the site administrator wants them to talk about, or whatever the site administrator lets them talk about. The web traffic turns directly into cash when a certain percentage of the site users click on advertisements. Even more important than actual revenue from advertising, usage statistics for sites become a key factor in establishing the market value of web companies. Aside from direct revenue, the site administrator can also boast about the size and scale of the site. The term "stickiness" is an online marketing term. The more a site sticks to the user, the better. Users keep coming back to sticky websites.

In this case, an online community makes your content stickier; people just cannot get it off of them.

The discussion forum becomes a means for a site to draw traffic and to show up more prominently in search results. As Adrian and Kathie Kingsley-Hughes explain in the 2006 book *vBulletin: A users guide* "A discussion board quickly creates a pool of knowledge on your site" (Kingsley-Hughes & Kingsley-Hughes, 2006, p. 6). This pool of knowledge "means that you get entries in search engines for a wide variety of topics and search criteria, which will very effectively improve your overall ranking" (p. 6). So all of that discussion means that your site shows up for more search terms. Given that the goal is to generate traffic to the site, the discussion forum becomes an easy way to generate more traffic for your site.

Gone was the high-minded and utopian rhetoric of belonging, of spiritual and emotional connections in cyberspace. As online community was refined into a feature set, a bundle of scripts that create discussion threads on a screen through interaction with a database containing tables of users and their roles and tables of discussion text, the utopian language of the previous era faded away.

Default ranks and post count. The discussion boards in each of these platforms looked like many different kinds of discussion boards that came before them, but the new systems brought a series of methods by which users were ranked and ordered.

A theory of motivation is baked into the design of systems for ranking users. In a section called "manage ranks," David Mytton explains that inside Invision Power Board, "Ranks are used to encourage members to post" (Mytton, 2005, p. 96). Similarly, in phpBB, "Ranks are a way of giving your community members a credit for their participation" (Stefanov, Rogers, & Lothar, 2005, p. 137). The token names that show up next to users' names on each of their posts are a form of encouragement, meant to communicate their status. Mytton explains exactly how this system works: "Once a member reaches the Min Posts value, their rank will be automatically changed. 'Pips' (or a custom image you can create) will appear beneath the member's name in their profile and in their posts at this point." He goes on to explain: "As their post count increases, they will rise up the ranks, and gain status within your community." Here again the idea of "status in your community"—the notion of reputation to others—is turned into something that is easily measurable inside the logic of the transactions on your website.

The more that individuals post, the higher their post count, the higher their displayed rank and, as such, the more status they have. "Used well, they encourage more posting and therefore greater activity" (Mytton, 2005, p. 96). Again, the goal of your community is to create activity, as that is what you are seeking as a website owner. As the site administrator, "You can define your own member ranks from the Manage Ranks option" (Mytton, 2005, p. 96). With that noted, the software comes with a set of descriptions for each of your community members. "From a fresh installation, you will see three already defined: Newbie (0 posts) Member (10 posts) Advanced Member (30 posts)" (p. 96). Straight out of the box, Invision Power Board has an idea of who your members are and how to categorize and motivate them to do the things that the software's creators think motivate the end users of anyone running the software. Underlying all of this is a theory of value. The ultimate goal of an online community is to generate value in the stickiness of users returning to visit the site again and again.

The logic of permission and control. Users and posts are the primary entities that these platforms are designed to manage. While one might initially think that the content of discussion—the posts organized into discussion threads—are the primary focus of what forum administrators moderate and filter, there is an entire system behind the scenes that organizes, categorizes, and controls the users. As Mytton explains, Invision Power Board lets you organize your users into categories. He explains that this is "mainly to assign permissions and thus control what your members are able to do on your forums" (Mytton, 2005, p. 97). The logic of these systems is one of permission and control, of deciding what "your members" can and cannot do. For example, in the Invasion Power Board system a moderator can select "yes" or "no" in the admin interface to any of the following questions regarding the abilities of individual users or particular categories of users: "Can mass move topics? Can mass prune topics? Can set topics as visible and invisible? Can set posts as visible and invisible? Can warn other users? And can use topic multi-moderation?" (Mytton, 2005, p. 89). Selecting "yes" for any of those options changes the user interface for that category of users. For example, with the ability to "mass prune topics," a user has access to a new capability of deleting posts en masse, based on particular criteria that the user applies. Each of these options describes a power or a permission that the admin can set the moderator up with—the ability to prune and reorganize the topics and discussions en masse and to ban users. In short, the entire system is available to the admin to set up what can and cannot be done, to control what has and hasn't been said.

The same vision of enabling different tools for moderation and control of posts through moderation and management of the privileges of different classes of users holds true in phpBB, where "a moderator is a community member who has some more privileges over the postings and the topics than those that the regular users have" (Stefanov et al., 2005, p. 57). Those privileges are offered to the extent that they "make the forum a better place for discussion in line with the administrator's vision for the community" (Stefanov et al., 2005, p. 57). The term "privileges" in these cases is important because of what it implies about the logic of permission and control. User capabilities and actions in online community platforms are framed as advantages or favors granted by site administrators. The discussion board is created, maintained, and controlled by granting particular privileges to users by the owners of a given online community.

As online communities began to hit their stride, a new phenomenon was taking shape. The idea of online community was, for many in the web application business, becoming passé. Technologists began to talk in terms of "social." Exploring the emergence of social networking and social applications brings to the surface disagreements about whether or not "social" had replaced "community." To some, "social" is the next phase in the evolution of online community. To others, "social" was something entirely different from online community. The disagreement about what the terms "social" and "community" mean in these technical books offers a vantage point to better unpack the ideologies at play in their definitions.

From community to social. The success of social networking sites such as MySpace and, later, Facebook, has shifted interest from online communities as places to social networks as assemblages of links between people and digital objects. The idea of "social" that came into web parlance with social networks and applications such as Flickr, Twitter, Yelp, and any number of other social sites is a primary concern of books about designing and managing online communities in the latter half of the last decade. Some of these books insist that "social" is simply the further development of online community, while others insist that it is something distinct but related. In either case, these divergent perspectives offer insight into underlying ideas and values at play.

Online community continues to refer primarily to websites that function as discussion forums and to time blogs that have extensive commenting and discussion components to them. The online community software platforms (phpBB, vBulletin, Invision Power Board, etc.) all continue to be widely used and increasingly represent a small niche genre of websites in the face of the continued massive growth of sites and applications built around the notions of social, such as Facebook and Twitter.

Why community became passé for some. On the first few pages of the 2009 book *Designing Social Interfaces: Principles, Patterns, and Practices for Improving the User Experience*, Christian Crumlish and Erin Malone explain why the term "social" has replaced the term "community" on the web. "In the early days of the Web, social experiences were simply called community and generally consisted of message boards, groups, list-servs and virtual worlds" (Crumlish & Malone, 2009, p. 5). They explain that these were "place-centric" gathering places and that there "was little distinction between the building of the tools to enable these gatherings and the groups of people who made up the community itself." They cite SixDegrees.com from 1997 as the first site "that straddled the line between community and what we now call social." They explained that "SixDegrees showcased connections among people, allowed users to create and manage their personal profiles, and brought people together based on interests and other features. Sound familiar?" The primary difference the authors are articulating is between the idea of virtual places in which people come together around shared interests and a notion of the individual user as a node, as an object, that is networked and connected to other users based on features of their profile.

Beyond becoming an outmoded notion, as Crumlish and Malone explain, "before the dot-com bust, *community* became a dirty word—most likely because it was overly resource-intensive to build and care for, and no one figured out how to make money from all that work" (p. 6). Instead of trying to create a community, instead of identifying an interest and attracting a set of individuals who might develop into a community around that interest into a set of people who could establish shared history, the logic of social is based on getting users to bring their real-world networks of relationships into your application. As the authors explain: "These sites count on each person bringing his personal network into the online experience" (Crumlish & Malone, 2009, p. 7).

This tension between community and social is resolved differently in other works. In the 2009 *Online Communities Handbook: Building Your Business and Brand on the Web*, Anna Buss and Nancy Strauss explain that because online communities are "websites where user relationships develop...social networks are online communities in their purest form" because "user relationships are the main focus and activity" (Buss & Strauss, 2009, p. 16). In contrast, in the 2010 *Design to Thrive: Creating Social Networks and Online Communities That*

Last, Tharon Howard, commenting on Buss and Strauss's claim that social networks are online communities in their purest form, suggests that "Web 2.0 biased definitions do violence to online communities, which have been around and accepted since the 1980s" (p. 13). For Howard, equating social networking sites with online communities erases many of the most important components of what makes online community powerful.

For Howard, the key difference between online community and social networking sites is that "social networks put individuals at the center of relationships" (p. 13). That is, the "the organization of relationships in social networks is fundamentally different than online communities." As an example, "In social networks such as LinkedIn or Facebook, the *individual* user is the center of the network" (Howard, 2010, p. 13). In these social networking sites "the view of the network which I see and the basis upon which connections are built is *entirely unique to me*" (Howard, 2010, p. 15). In contrast, "communities are different." In online communities, Howard suggests that one's relationship to other users is secondary. Further, "The individual makes a commitment to the group as a whole before other members, but paradoxically, this commitment means that the contacts that I have with other members of a community is richer, more complex, and more predictable than the contacts I can make in a social network" (Howard, 2010, p. 15). As a result, Howard suggests that "online communities can achieve the kind of complex organizational structures needed for both cooperation and collective action; social networks can't" (Howard, 2010, p. 22).

In short, the definitions of social networking sites and online communities are overlapping and contested. With that said, the differences in how Crumlish and Malone (2009), Buss and Straus (2009), and Howard (2010) define and reflect on the concepts of "community" and "social" illustrates the edges of the disputed definitions. Social networking is constructed on the idea of building sites and web applications around users and their relationships. When users log onto a site like Facebook or LinkedIn, they see a completely customized site based on the activity of those they are connected to. In contrast, everyone who visits a site running phpBB or vBulletin, or for that matter a Google Group, sees largely the same set of discussion areas and topics and posts. Crumlish and Malone, Buss and Straus, and Howard each categorize sites like Facebook and LinkedIn as social networks and the discussion platforms as online communities. The difference comes in evaluating the relative value of the online communities. Of the three books discussed in this section, only Howard's argues that there is something particularly important that is lost when we move from the place-centric common discussion space to the user-centric view of activity from one's connections, friends, or contacts. For Howard, the shared experience of participating in a single ongoing hub of activity built around reading and responding to the same ongoing discussion is critical. In contrast, he suggests that social sites are far more fractured in that they are built around public interactions between users instead of sustained common discussion.

The fact that online communities haven't disappeared in the hype of "social" suggests that there is credibility to Howard's perspective. While in the minds of those developing social web applications online community may have become passé, there is still an extensive set of thriving discussion boards across the web. For example, the anime-themed Gaiaonline site (built on phpBB) sees over 1 million posts a day, is visited by 7 million unique users a month, and boasts more than 26 million registered users. Similarly, the continued refinement and popularity of discussion sites such as the ever-expanding network of Stack Exchange question-and-answer sites illustrate how these community-centric models that downplay social relationships in the interest of convening and focusing attention on discussion and deliberation suggest that Howard is right to argue that there remains a value and interest in this place-centric notion of online community. Nevertheless, these community-centric forum models have become a niche genre of website. Facebook now boasts over 1 billion users, and Twitter boasts more than 400 million daily posts from users. Beyond the massive scale of their use, both Twitter and Facebook's recent initial public offerings illustrate how the financial markets value this vision for the future of social over the notion of community.

Community, social, and control. Despite disputes about what constitutes an online community and a social network, there is considerable continuity in how Crumlish and Malone (2009), Buss and Straus (2009), and Howard (2010) see the goals of both. As designers of a social network or an online community, the owners' objective is to get their users to do what they want them to do. In this section I briefly explore the emergence of reputation systems, consideration of user agency, and how users become a disposable commodity in the language of these three prominent and recent books on designing and running online communities.

Emergence of reputation systems. More and more, the design and structure of reputation systems has become a central feature of books about online communities. Where platforms like phpBB and vBulletin include relatively simple systems for assigning ranks to users, the new social networks implemented in various social and community platforms have become far more sophisticated. The design and function of reputation systems will be discussed much more extensively in a later chapter six dealing with their functional role. With that said, it is useful to dig a bit into the theory of reputation systems here to understand how discussions of online community establish a logic of control.

Crumlish and Malone offer the following explanation for the value and importance of reputation systems.

> Including reputation metrics and services in your social interface is somewhat less ambitious than trying to measure people's real-world reputations or even trying to capture their online, virtual reputations. You can simply focus on the communities you are fostering in your application, the values you are trying to instill in the environment, the behaviors you wish to encourage, and the types of people you wish to engage ever more deeply in your social environment. (Crumlish & Malone, 2009, p. 153)

This description is considerably different from the ideas of community in the 1990s. Instead of hosting, convening, or facilitating online communities, Crumlish and Malone are interested in what many on the web are calling social engineering. Their suggestion to designers of social interfaces to step back and think about the values they want to instill in their environment, the behavior they wish to encourage, and the types of people they want to engage is focused on getting people to perform as you, the designer, want them to perform. This isn't about designing the facilities for communities to happen in; it's about literally designing social interaction. This has become a process of monitoring, tracking, and modifying the behavior of individuals through application design. For these authors, designing the reputation system for a social interface is a process of behavior modification. Crumlish and Malone are not naive on this point. They suggest that "each designed model of participation and reputation embodies its own set of biases and incentive structures" (2009, p. 154). They are not simply designing the information architecture of a website; rather, they are designing models of participation and reputation that represent a social architecture of relationships.

The logic of behavior modification is evident in how Crumlish and Malone suggest designers conceive and structure their reputation systems.

> Fundamentally, a reputation system involves tracking desirable behavior and then recognizing it publicly. So, any well-designed reputation system is going to start with an inventory of desirable behaviors. Do you want to make sure people try out a certain

feature, or strive for higher-quality contributions, or log endless hours responding to others? Just as managers say, "You can't manage what you can't measure," the same applies to reputation. You can't acknowledge what you aren't tracking. (2009, p. 154)

Crumlish and Malone start with an inventory of intended behaviors. What do you want people to do? How can you measure if they are or aren't doing it? From there, the goal is to simply poke and prod users into doing what you want them to do.

The feeling of agency. Given that Tharon Howard (2010) had such a different vision for the power that online communities could provide over social networks, one might imagine that he would articulate a fundamentally different notion of how control and power should work in online communities. At certain points it seems like this is the case, but a close reading suggests otherwise. Consider the following description Howard offers of users' sense of control:

> How do you help your individual members feel they are in control or have influence over their environment and yet still achieve a balance between meeting the needs of an individual on the one hand and protecting the goals of the community on the other? (Howard, 2010, p. 81)

What originates in a caring for user agency is fundamentally not about putting members in control but creating the illusion of control. Community designers want to meet users' needs, but they also want them to help you meet the needs you have identified for the community. In this sense, Howard's conception of control is very similar to the patterns of control documented thus far. In this story, the owner/designer/site administrator of the community or social network wants to make members feel as though they have power. Feelings—or possibly illusions of control—are in Howard's perspective an important thing to build into online community systems. He offers a similar point a page later. One of Howard's central points is that users need to have influence in the community. This suggests that he wants to cede control to users, that he might be interested in notions of how communities develop their own governance. However, his definition of when members have achieved influence in a community belies a very different vision. For Howard,

> Influence can be said to exist in a community when its members believe that they can control or at least shape the policies, procedures, topics, and standards of evidence used to persuade others in an online community or social network. (Howard, 2010, p. 82)

The key word in this sentence is "believe." Just as Howard wants to make sure members feel they are in control, he also wants them to believe they have a voice.

The first line of the foreword to Howard's book is telling. "Communities don't happen by accident, although you'd never know it from the haphazard way most companies go about trying to create them" (Howard, 2010, p. ix). Within the context of the book, communities are things that companies create. Communities are not accidental or haphazard. They are manufactured and designed to get particular kinds of people to exhibit particular kinds of behaviors.

Using up your users. A chapter on motivation from Anna Buss and Nancy Strauss's 2009 book *Online Communities Handbook: Building Your Business and Brand on the Web* is similarly useful for exploring the interplay between a set of psychosocial beliefs that underpin the authors' explanation of desired functionality in online community software. From the beginning, it is clear that not only is there a commercial goal for the sites, but that site users are explicitly thought of as commodities themselves. For example, when Buss and Strauss explain that the key question is "How can you get the maximum mileage from your members?" the community members explicitly become a fuel to be consumed (Buss & Strauss, 2009, p. 80). Similarly, in a heading for a section on how to engage members, the authors suggest that you "Grab them when they're fresh." Thus, members become something akin to a perishable fruit or vegetable. In both cases users become something for you, as the site administrator, to consume. The authors go on to explain how this approach translates into specific functionality.

For Buss and Strauss, the first activity on your site needs to be simple and enjoyable, show "an immediate result (for example, a photo appears on the page) for instant gratification," and promise a long-term result, such as another member's comment on it. The reference to "instant gratification" itself represents a longstanding connection between the discourse of marketing and the discourse of behaviorist psychology. The extent to which this represents a rationalization of functionality on the terms of marketing is unclear. Likewise, it is questionable whether this is indicative of an ideology that has itself guided the development of these kinds of gratification cycles in the functionality of online community software. In any event, the link and relationship merits further study and exploration.

At this point, Buss and Strauss discuss the marketing term "stickiness" that has already come up in the books studied and has become a key term in eCommerce. Buss and Strauss describe stickiness as "website content that causes the user to spend more time on the site" (2009, p. 81). Stickiness is grounded in their pop-behaviorist psychology principles as an explanation for how to run an online community. As in much of the book, the psychological ideas and the theory of society and community embedded in those notions are intended as statements of fact. These are simply statements about the way the world works that the authors translate into ways of understanding how to structure online community. That is, Buss and Strauss do not provide an explanation of how their vision for running online communities is based on a particular behaviorist set of assumptions about psychology. The behaviorist perspective is an underlying assumed value. As they explain

> Not only do you need users to spend time on the website, you need them to come back again and again. You should therefore aim for the right balance between instant and delayed gratification. On the one hand, users should see instant results from their actions, giving them a satisfying experience on the website. On the other hand, there should be benefits that build over time, enriching the experience the more time they spend in the community. (Buss & Strauss, 2009, p. 83)

Buss and Strauss mobilize a theory of human behavior that is focused on instant and delayed gratification in service of particular designed features. The functionality they describe is a staple of the design of online community systems. The authors explain that loyalty programs in which "users earn points for each website action" can be valuable because they can frequently encourage users to return to the site by serving their "self-gratifying desires" (Buss & Strauss, 2009, p. 83). The authors contend that "virtual rewards" trigger both instant and delayed gratification. They remind us that these rewards should "depend on your target demographic" and suggest that "teenagers are not interested in a business card exchange feature, while business users may be less likely to crave virtual pets to keep on their personal homepages." The authors' theory of self is based on behaviorist models. Reward the behavior you want, gratify the user, make and satisfy their cravings, keep pulling them back in and providing the stimulus of the reward.

The features Buss and Strauss discuss have more recently been described as "gameification." "Gameification" claims to use principles of game design in non-game situations, although ideas about points, badges, and virtual rewards clearly predate the notion of gameification. Margaret Robertson's critique of gameification as "pointsification" (2010) and Ian Bogost's critique of it as "exploitation-ware" (2011) are well-argued critiques of this type of thinking about users.

The behaviorist psychology of gratification and rewards becomes a theory of society as Buss and Strauss begin to explain the critical value of "social hierarchy" as an explanatory device for another particular set of functionality—the reputation system. They provide us with a theory of the social and then give us a case study from a site they worked on (Ciao.com), in which they illustrate what this theory looks like in practice. It is worth quoting both of these at length to pick apart exactly what the authors are suggesting and what their vision of a social hierarchy looks like as a system.

> Social hierarchy in your community is a powerful tool. Just like offline communities, online ones quickly sort themselves into a hierarchical structure, normally with the most experienced members at the top. By encouraging hierarchy in your community and offering visible status symbols based on seniority and activity level, you can create an environment in which members feel as if they are working toward an objective: the next rung on the social ladder. (Buss & Strauss, 2009, p. 86)

Here Buss and Strauss explain that one of our most powerful tools is not part of the software; it is social hierarchy and our innate desire for social hierarchy. When the authors assert that online communities "quickly sort themselves into hierarchical structures," they establish a social framework that informs their approach to design. Their assertion that this is "just like offline communities" makes this out to be inevitable. They go on to suggest how this vision of human behavior can inform the design and development of software. For example, the authors recommend that a site admin will want to offer "visual status symbols" of the characteristics they want to encourage—in this case "seniority and activity level"—and give users the feeling that they are "working toward an objective." Specifically, the members' objective is to reach "the next rung on the social ladder." The instant and delayed gratification in this system are based explicitly on the idea that what motivates users (again the title of this chapter is "Motivation") is their desire to gain some arbitrary and non-monetary signifiers of their increasing social status in the given community site. Buss and Strauss go on to explain:

> The consumer community Ciao.com has a non-monetary rewards system that issues colorful dots as a status symbol. Members can earn points to change the color of their dot by posting product reviews that other members rate as useful, or by performing other community actions. You can find many members who post messages on their profile pages related to this community points systems. "Hooray, I'm finally red," they

write. "Please read my product reviews and help me turn orange!" (Buss & Strauss, 2009, p. 87)

Buss and Strauss begin by describing Ciao.com as a "consumer community" and explicitly suggesting that the community uses a "non-monetary rewards system." Again, the very idea of a "rewards system" brings with it the behaviorist inclinations of gratification. At each step of the description we see ways in which users are trivialized. As users engage in the activities that Buss and Strauss have chosen to reinforce—in this case writing product reviews and performing other undisclosed "community actions"—they receive the points that enable them to change the color of their apparently arbitrarily and infinitely trivial "colored dots." They tell us how their users exclaim "hooray" at finally being red and attempt to recruit each other to read their product reviews to "help me turn orange!" In Buss and Strauss's presentation, users and community members are commodities that can be controlled based on their desires for gratification. The structure of social interaction enacted and prompted by the software's reputation system and use of virtual rewards is anchored in ideas about users' innate desire to climb the social ladder. The functionality of the software reifies a set of ideas about social hierarchy in the minds of the authors as a set of features with which users directly engage. The somewhat abstract notions of social relations become concrete in the way that they are operationalized and instantiated in the functionality of these systems.

The alternative age of participation. There is an alternate vision for what online community can be. In the 2009 book *The Art of Community: Building the New Age of Participation* author Jono Bacon offers a fundamentally different perspective on online community. Published by O'Reilly Media, this book does not focus specifically on online communities as much as on how to form and support communities of people working toward common goals. As the community manager for Ubuntu, one of the most popular distributions of the Linux open-source operating system, Bacon brings a distinct perspective to online communities. In contrast to just about everything else being written about online community during this time period, he focuses on how a range of free culture projects and initiatives convene and support a community of members to work together and ultimately take on their own governance.

To illustrate the kinds of free-culture movement communities he discusses of people who band together to work toward some ends, Bacon offers the example of a set of people who came together to create a free and open-source audio editing application. "We set up a code repository, a website, mailing list, and a bug tracker, and scheduled regular meetings. We organized hack days, bug squashing parties and online discussions to plan and decide on major architectural decision" (Bacon, 2009, p. 23).

In the process of developing this open-source audio editing application, the communication tools—like the mailing list and the website and the code repository—are employed as tools to support the work of the group. The communication tools are not discussed as a set of features that define the community or as a carefully assembled system of behavioral modification.

Read-mostly and write-centered communities. Nearly all of the communities, discussion forums, and social networks discussed in some of the previous books fall into what Bacon would call "read-mostly communities." Bacon, however, is primarily interested in "write-centered communities" (Bacon, 2009, pp. 34–35). The difference here is not one of reading and writing text, but of accomplishing work and having a stake in shaping the goals and objectives of the community. The author is borrowing from the language of computer storage systems, wherein writing to the drive changes its contents and reading from the drive simply presents information.

For example, he describes a "read-mostly community" as something where "a group of fans provide feedback on a forum." In contrast, a "write-centered community" is something like the "free Culture communities such as Linux, Wikipedia, OpenStreetMap, Creative Commons, etc." (Bacon, 2009, p. 36). In these write-centered communities, "Collaboration goes much further. It becomes much deeper, more intrinsic, and more accessible to all. Instead of merely *enjoying things together*, collaboration goes so far as to help people *create things together*" (Bacon, 2009, p. 36; emphasis in original).

Shifting from enjoying "things" together (such as particular pieces of music or bands) to collaboratively creating them is the central idea in Bacon's distinction. To be sure, it is actually far more difficult to separate out the "enjoying" things together from "creating" things together. A fan-fiction web forum (a forum in which fans of a TV show or novel write and share their own show-inspired stories) is an example of individuals creating the objects being enjoyed by the community. To Bacon's point, examples such as fan fiction sites are different from sites like Wikipedia or OpenStreetMap. In those sites, a group of volunteers bands together to design, administrate, and govern the ongoing production and maintenance of a free culture product and/or service. Acknowledging that there are shades of grey between enjoying and making things is not to concede the point that the networks of volunteers who band together to create and run things like Wikipedia or OpenStreetMap are distinct from discussion-centric online communities.

The core vocabulary that Bacon uses to describe community—particularly in the write-centric, free culture communities—is fundamentally different from that of just about all the other contemporary web books. While this puts him in a minority position, it's valuable to interrogate his perspective to show how stark the contrast is between his view and the broader contemporary perspective and to suggest a vision for what a different language and vision for online community might look like.

Enable. One of Bacon's key terms—"enable"—shows up repeatedly in the book. Here he describes why that term is so important to his conception of community.

> I have tried to summarize what we community managers do in one sentence. The best I have come up with is: I help to enable a worldwide collection of volunteers to work together to do things that make a difference to them... Twenty of those twenty-one words are really just filler around the word that I *really* think describes what we do: *enable*. Our function as community leaders is to enable people to be the best they can be in the community that they have chosen to be a part of. Our job is to help our community members achieve their greatest ambitions, and to help them work with other community members to realize not their own personal goals, but the goals of the community itself. (Bacon, 2009, p. 14)

Bacon is interested in recruiting a network of volunteers to work together toward common goals. The word "volunteer" is similarly important in this context. Bacon doesn't think in terms of users or participants, but rather volunteers. A volunteer is distinct from those other notions of membership in that it comes with the sense that the person is acting in an altruistic fashion to promote a public good. The vision of online community Bacon is interested in (and is promoting in his book) has much more to do with community organizing as a form of social activism than it does with the bulk of the literature on online communities.

Governance. Two books published in or before 2000 discuss and explore the importance of the notion of governance in online communities. There is only one post-2000 book in the collection of online community books I have pulled together that devotes serious attention and consideration to governance. Chapter 8 in Bacon's *Art of Community* devotes more than 50 pages to the topic. Other books ignore the topic, and by doing so imply that online communities are, in terms of functionality, dictatorships or company towns. Where many of the other authors may be interested in the appearance or feeling of control in the hands of the participants, the majority of the authors of books considered in this chapter are focused primarily on extracting value from the users or members to serve the desires of the owners.

Bacon begins his discussion of governance by describing a particular web phenomenon, the Ubuntu forums. (These forums are run on the vBulletin software, discussed previously.) However, it's not the features of the forums or the way that moderation tools in the software work that is of particular interest to Bacon. He explains that there are moderators in the forums who report to a Forums Council, who in turn report to the Ubuntu Community Council. Outside any definition of roles or permissions within the vBulletin software, Ubuntu has established a system by which members of the Ubuntu community have explicit roles to play in the governance of the project.

40 years of online community. A lot has changed in the 40 years since the mimeographed brochure for The Community Memory proclaimed the creation of "strong, free, non-hierarchical channels of communication—whether by computer and modem, pen and ink, telephone, or face-to-face" as "the front line of reclaiming and revitalizing our communities" (as quoted in Bell et al., 2004, p. 14.) The resulting history has given us a series of platforms and a range of visions for what online community can and does mean.

Community as functionality and as product. When describing platforms such as phpBB, vBulletin, and Invision Power Board, the authors of these books describe community as a set of features. An online community came to be a set of system-defined roles and the rules that govern what different kinds of users can post and who can moderate. At the same time, other authors interested in more sophisticated definitions of community— specifically, as something that emerges as a result of that feature set—were transforming the idea of designing the features and functionality of systems into designing and architecting our actual relationships to each other as parts of those social systems. For the most part, voices in these books that pushed for thinking about community as something "sacred" or as something that shouldn't be perceived as being owned, bought, and sold are few and far between.

Persistent free information/open source undercurrents. There is an undercurrent throughout the history of online communities that is entangled with the values of the free software/open-source software movement. The very first book considered in this analysis, Bowen and Peyton's *The Complete Electronic Bulletin Board Starter Kit* (1988), situated BBSes in the free software movement, and Jono Bacon, community manager for Ubuntu and author of

The Art of Community (2009), similarly situated his vision and values for online community in that movement.

Increasingly sophisticated notions of control and manipulation. Even in the earliest books in this analysis—books about BBSes—sysops were making decisions that kept people with 300k modems off their boards because they thought cheaper modems marked them as undesirable. Since those days, books about designing, running, and managing online communities have presented an increasingly sophisticated set of ideas for how you can get people to do what you want them to do in your online community while providing an illusion of agency. The next chapter in this book develops and documents how that control and manipulation is enacted.

Better and worse kinds of dictators. To acknowledge that online communities are, for the most part, fully controlled and operated by owners—something akin to dictatorships—is not to suggest that they are inherently bad or wrong. This is just to acknowledge and appreciate online communities for what they are and to recognize that, in this sense, they are very different from what we imagine to be the case with community in general. Recognizing that this is how control works in online communities poses a new question: How do the members of an online community feel about the particular dictator in this particular community? That is, some of the modes through which online community is controlled and enacted resemble the paternalism of a benevolent dictator. In these cases, community members are being manipulated and nudged to act in particular ways, but that tactic is more or less transparent, and they are willing participants in something that ideally has their best interests at heart. In contrast, there are other modes in which the dictatorship is exploitive, in which users are less aware of how they are being manipulated and are in effect being taken advantage of and exploited.

· 6 ·

ENACTING CONTROL, GRANTING PERMISSIONS

The history of the rhetoric of online communities suggests that they are, in general, products and commodities. Online communities are targeted toward particular ends and are run and administrated toward those ends. Which brings us to a few related questions: How exactly is control of online community enacted? To what extent is that control absolute? Assuming that control is not absolute, how do members and participants in online communities exercise agency, and how do the owners of online communities maintain control in the face of attempts at user agency?

I have organized this section according to three different technical modes by which the creators, owners, and managers of online communities exert control over these communities. I've called these (1) visual design/information architecture, (2) moderation tools, and (3) reputation systems. Aside from identifying the modes of control, I am also interested in identifying the actors who use these modes. In this case, control is enacted through a distributed network of designers, site administrators and moderators, the site's users, and routines in the site's software that make use of underlying networking protocols. Control of online communities is achieved as these actors operate through the three modes. The description of tools and tactics for control bring with them anticipations of the reactivity of members. That is, we can come to understand the agency of community members in the strategies and tactics the owners of the community mobilize.

Visual design and information architecture. The design of a site is intended to encourage and discourage particular kinds of behaviors in users. These include both overt signals about the kinds of people that are invited to participate—such as a site's tagline explaining that it's a "place for women"— as well as more or less overt approaches to shaping the nature of dialogue and discussion in an online community, such as how prominent the "post" or "comment" button appears on a site. This mode of control is primarily the domain of the designer. In some cases, this will be the web developer who created the particular site. Much more often, however, the role of designer is actually itself distributed between the individual who configures and localizes the look and feel of a particular community software application (such as phpBB or vBulletin) and the entire group of developers who have designed the system they are using.

Getting the right tag line. The most visible level at which designers effect the shape and nature of interactions and discussion through visual design and information architecture includes things such as the name, logo, and tagline for the site. Discussion of how to choose these things, as well as the value in making sure your site is distinct and stands out, shows up in most of these books. Remember the earlier discussion of Amy Jo Kim's commentary on the tagline for L'Eggs pantyhose site? She suggested, "Imagine, if the L'Eggs community had used a tag line like iVillage's 'Real Solutions for Women,' their site would have evolved in a very different way" (Kim, 2000, p. 22). The implication is that they could have done a better job attracting the women that the owners of the site wanted to attract, while dissuading the pantyhose-wearing men they actually did attract if they had had a better tagline signaling who was welcome and who wasn't. In this sense, the design of a site serves to signal who should and who shouldn't participate. This point is reiterated through a range of more subtle features of community sites than their taglines.

Bury the post button: Page layout, empowerment, and manipulation. Derek Powazek's *Design for Community: The Art of Connecting Real People in Virtual Places* (2002), focuses on how to add what he refers to as "community features" to websites. One of the focal points he stresses is how the design and layout of a site should structure user participation. The excerpts that follow come from his section "Rule #2: Bury the Post Button."

In my experience with community features, I have observed a proportional relationship to the distance that the post button is from the front door of the site and the
quality of the conversation on the site. The farther away it is, the better it gets.
(Powazek, 2002, p. 53)

The post button is intentionally placed to provoke a specific kind of conversation from a particular kind of user. At the end of his first sentence we
find what Powazek sees as the value that "community features provide." In his
case, the value he is trying to optimize is the "quality of the conversation,"
something that he will further explain shortly. With that goal in mind, he
offers a theory of visual design that will strike many as counter-intuitive. To
explain this idea, he coaxes us to think about the metaphorical "front door" of
the site. Like many of these texts and works on web design in general, Powazek
uses the layout of a building as a way to describe the experience of moving
through a site. In doing so, he—along with many of these texts—spatializes
sites. What was a series of files, or a series of linked documents, becomes a
home or a building. One can read into this the implication that we aren't
having intimate or quality conversation on the front porch or in the parlor.

With this said, if the goal of a site is to engage in dialogue, why would
one want to "bury the post button"? The point goes against much of the common wisdom of web design exemplified in works such as Steve Krug's *Don't
Make Me Think: A Common Sense Approach to Web Usability* (2000). Why is
Powazek suggesting the obfuscation of functionality? He goes on to explain:

Why would this be? Because, in this case, the multiple clicks it takes to read the
whole store are actually a great screening mechanism. Users who are looking for
trouble or aren't really engaged in your content will be put off by the distance. They
lose interest and drift away. But the users who are engaged by the content and interested in the results of the conversation will stick with it. These are the people you
want to retain, because they're much more likely to post great thoughts. (Powazek,
2002, p. 53)

Making it more difficult for users to get to the post button and to respond
and share their ideas is part of an explicit attempt to generate a particular kind
of discussion. Here we find out a bit more about what it is that Powazek considers quality conversation. It has to do with separating out different kinds of
users. Those "users looking for trouble" and those who "aren't really engaged"
are the kinds of people he wants to filter out. In the process, he hopes to retain
the "users who are engaged by the content and interested in the results of the
conversation" (Powazek, 2002, p. 53). In short, Powazek has in mind a set of

categories of good and bad users, and the design decisions he is providing are intended to result in a particular kind of discussion between the good kinds of users.

In this case, the design and relationship among pages, and making someone click though multiple pages, is the instrument, or the tactic, he uses to configure discourse. He goes on to explain that this "can translate into different things when applied practically." As an example he suggests that "perhaps the best place for the call to action ('Post your thoughts!') is at the bottom of the page instead of the top" (Powazek, 2002, p. 53). Notice here that the distance metaphor—that is, how far away from the front door the individual is—has now shifted from clicks to scrolling down the page. He explains: "That way at least your users will have had to skim through some content before they are given the chance to respond." So distance operates as both a function of clicks and of page layout. Both the design of the structure of a site with "community features" and the visual design of the individual pages is being explicitly presented in terms of structuring both of these features to create particular kinds of discourse and dialogue.

This may seem like a somewhat self-evident point; of course designers are designing according to their goals. However, this holds serious implications for what anyone who studies conversations and discussions in online communities can say about what the textual records on a page of a particular online discussion can tell us. Any interpretation of online discourse needs to start with the recognition that, in all likelihood, the site has been designed to invite and engage particular kinds of people in particular kinds of discussion. That said, and given that decisions about what tools to use to run a site are made without a full understanding of the design, limitations, and structure of the underlying software, there is a good chance that the software is also at odds with what a given community manager wants. The intentions of the software creator can be at odds with the will of the administrator.

Here the specifics of this description become important. One suggestion for researchers is to consider directly the point Powazek proposes: that this is a particular design tactic, and that if one wants to engage in a study of online discourse, it is probably a good idea to look at where the post button is, since the placement and location of the call to action is likely to act as a filter. Understanding exactly how that filter works and who is being filtered out or in is always going to be a tricky game. However, it is essential to realize, at the base level, that the designers of sites are using visual design and information architecture in an attempt to prompt particular kinds of people to particular kinds of actions and discussions.

Amy Jo Kim offers a similar perspective in her 2000 book *Community Building on the Web: Secret Strategies for Successful Online Communities*. She explains: "What you want to do is create appropriate hurdles for contributions" (Kim, 2000, p. 71). In this case, "It's up to you to figure out the restrictions that best meet the needs of your members and support the kind of community you are trying to create" (Kim, 2000, p. 71). The visual design of a given site can be organized to create exactly these kinds of hurdles and barriers that serve to limit contributions to one particular type.

Talking like the right kind of person. Under the heading "Talk Like a Person," Christian Crumlish and Erin Malone suggest in their 2009 book *Designing Social Interfaces* that "savvy enterprises appreciated the value of communicating to potential customers in a human voice" and go on to explain that "the corporation has always been a mask that disguises the human nature of the people who do the actual work of the business" (Crumlish & Malone, 2009, p. 26). They suggest that site designers attempt to write the text for all parts of the site (from the text in the sign-up process to the text in the error messages) as you would talk to them like a person and not like a computer system.

"But what kind of person?" they ask. "The type of person you hope will get involved in your site," they answer. They go on:

> Model the sort of tone and personality you're aiming to recruit. This is all the more true in the context of social sites. If a website does not communicate from the get-go that it is populated, and written by, ordinary human beings, how will people ever feel comfortable there? The antiseptic air of a hospital or the bureaucratic formality of the Department of Motor Vehicles is no environment for fostering connections, relationships, or collaboration.... The bottom line is *authenticity*. Would you really say that? Can you read it out loud without cringing? Does it *sound* like your kind of people? (Crumlish & Malone, 2009, p. 27)

Following this suggestion involves creating a theory of the user in the mind of the developer, the idea being that the software they are writing is in effect a stand-in for themselves talking to the user. Just as the placement of the buttons for response is intended to nudge particular kinds of users into particular kinds of actions, the text of a site is likewise crafted to create an air of authenticity. Crumlish and Malone go on to explain exactly why you want to do this. According to them, having your site "talk like you," and having it talk to people whom you imagine like you, puts your users in an important "receptive state of mind" that "permits the reader to enter into a dialogue with the site and reinforces the feeling that the site is made by people and not machines" (Crumlish & Malone, 2009, p. 28). While users are actually talking with a machine as they work through the sign-up process for a site, or when they hit an error message, they want the user to feel at ease in this receptive state of mind. The language and voice of the system that users interact with are intended to push them to forget that they are engaging in interaction with programmed rules of the application and to make them feel like they are engaged in dialogue with the developer.

Welcome to the site. Crumlish and Malone also suggest that you create a "welcoming area" for your users. The idea is to "Provide the new user with a warm gracious welcome to your site and services. This can be a special welcome screen right after registering for the service or a special email highlighting features" (Crumlish & Malone, 2009, p. 73). In this case they are suggesting a particular step in the information architecture and design of your site. The terms "warm" and "gracious" here are curious. What exactly constitutes a warm and gracious automated script that produces a set of texts on the web page? This advice makes sense according to the logic of Crumlish and Malone's suggestion that the site should talk like a person—specifically, that it should talk like they are talking to you. They want to impart a set of feelings and values in the language they use and the style it's presented in to make it feel warm and gracious.

Crumlish and Malone see this in terms of the kinds of welcoming events one might organize when someone comes to a new place. They suggest: "Providing a welcome area or start space is akin to orientation for a new job or college, or giving your friends a tour of your home the first time they visit" (Crumlish & Malone, 2009, p. 75). The site speaks for them, acting as a tour guide and speaking in the voice of its developers.

The sentiment here is similar to that of Cliff Figallo's perspective from a decade earlier, in his 1998 book *Hosting Web Communities*. Figallo suggested that "the interface works like a combination of your welcome mat, your front door, and your living room. If it works to bring visitors that far into your site, you have a chance of introducing them to the rest of what is there" (Figallo, 1998, p. 148). The structure and design of a site is established to prompt particular actions. The series of pages and orientation steps is designed to push them toward particular kinds of activity. Crumlish and Malone offer this analogy: "The more welcoming you are (in a light-handed fashion, of course), the more your users will feel comfortable and *want* to spend time on your site" (Crumlish & Malone, 2009, p. 75). Herein we see their ultimate goal: set up your site like this so that the visitor will want to spend time there. This has the effect of making them feel at home and happy on your site as well as realizing the marketing goal of "stickiness." That is, you want to make something meaningful so that people want to hang out on your site, and you make your money and establish the value of your application by getting a lot of people to spend a lot of time on it.

Moderate, filter, ban. When Stoyan Stefanov, Jeremy Rogers, and Mike Lothar introduced phpBB in their 2005 user guide *Building Online Communities with phpBB 2*, they present it as two things. In their words, "phpBB is a free, open-source Internet community application, with outstanding discussion forums and membership management" (Stefanov, Rogers & Lothar, 2005, p. 1). Aside from the forums, phpBB is a membership management system. At the heart of these applications—platforms such as phpBB and vBulletin, and the various social web applications—is a process of managing, moderating, and filtering users and their contributions.

Most online communities use technical mechanisms to moderate discussion and interaction. Sites run obscenity filters. Users can be banned, and discussions can be pruned, broken apart, hidden, reorganized, and edited. In most cases, moderation is enacted through the interplay between scripted rules in software systems and the judgment of site members empowered with the permission to use a range of moderation tools. Examining the nature of moderation in online communities offers insight into the interplay between established and enforced rules of etiquette and what is technically possible for a system, given the underlying features of network protocols such as HTTP.

In this section I first talk extensively about how moderation tools are configured in discussion board software. This provides a general overview of the kinds of things that can be done in any system and focuses on how these have been designed and configured in these online communities straight out of the box. It is helpful to look at applications such as phpBB, vBulletin, and Invision Power Board, as many of the more custom-built systems for other online communities are more particular and idiosyncratic. From there, I consider a series of historical developments in more generic design patterns that show up over time across more customized systems.

Understanding how and why moderation tools are used for particular goals provides an important part of the context for studying the records of online communities. It also offers insight into how communication in online communities is dependent on an interconnected set of procedural scripts enacted by software and a range of actions offered to different types of users with different levels of permission.

Finely tuned control of users: 10 types and 5 levels of permission in phpBB 2. When you visit a discussion board running phpBB 2, it looks relatively straightforward. You see a range of discussion areas, each of which has in it a set of threaded discussions. Once you sign up for an account you can respond to discussion topics and create new topics. What a user who is new to a system like phpBB isn't aware of is just how complex the same system looks from the site administrator's perspective. Different kinds of users with different types of permissions are able to see and use different functionalities.

The phpBB 2 software comes with ten permission types: view, read, reply, post, edit, delete, sticky, announce, vote, and poll create. It also comes with five different levels of permission: all, registered, private (a label individually applied to members who get access to a private section), mod (moderator), and admin (administrator). Most of these are self-explanatory: view, read, edit, and delete all control your ability to see sections of the discussion board and to make and edit posts. Sticky is an ability to make a discussion topic "stick" to the top of the section, so new discussion threads don't replace it as the first discussion in a given section. Announce lets you make site-wide announcements that show up in a text box across the pages of the site, and the voting and poll permissions enable users to create and vote in polls that show up as their own threaded discussions. The five permission types come with default sets of permissions.

The "all" level shows what users who haven't logged in can see. It's worth noting that in many cases there are sections of phpBB discussion boards that are not viewable to users who haven't logged in. In general, the web-harvesting and crawling tools such as Google are unable to log in to these sites to capture these parts of online communities. This means that (1) these parts of online community discussions do not show up in search results and (2) are not recorded in archived copies of sites such as those available through the Internet Archive's Way Back machine. These permission types are not unique to phpBB; they are more or less the same across tools like Invision Power Board and vBulletin. As a result of this design, anyone working with archived copies of discussion forums is very likely looking at a small fraction of the discussion content of these sites.

Becoming a registered user of a phpBB site is the basic level of entry for participation. At this level, the user can generally view most of the discussion board sections and has the ability to post, edit, and delete his or her own comments. Still, there can be other sections of the discussion areas that are visible only to members that an administrator or a moderator has designated as having "private" access, as well as other sections that are visible only to moderators and administrators.

The result of all these layers of permission and control is that there is likely always something that is obscured behind what we can see on the screen as particular kinds of users. For any of these systems, a table in a database ultimately establishes the level of permission associated with each individual account, one that creates a filter on the discussions and available options that a user sees in the discussion forums. Aside from what users can or cannot see, these systems include automated tools that moderators are enabled to use to manipulate communication in much more substantive ways.

Tools for moderators. Consideration of Invision Power Board 2's moderation tools as described in the software's user guide provides a frame of reference for understanding the kinds of things that moderators are able to do in these systems. In each of these cases, the logic of the tools of online community is exposed. A user with the right level of permission (in this case the moderator) is empowered to take particular kinds of actions on the content in the database and the processes that the online community's software runs to support communication. Ultimately, all of this functionality is dependent on the underlying constraints and designs of the database that contains information on posts and users and the kind of information that can be collected about a site's visitors and users through the underlying Hyper Text Transfer Protocol that enables those visitors and users to access the system.

The user guide for Invision Power Board 2 documents the full range of ways in which the system empowers moderators to manipulate and control discussion text. Nearly all of Invision Power Board's features are also available in tools such as vBulletin and phpBB and are explained in their respective guidebooks, so I will interweave some description of related features into this section. Briefly describing and explicating the different kinds of moderation tools these platforms come with provides insight into exactly how control is accomplished.

Inline moderation tools. The inline moderation tools of the Invision Power Board platform allow moderators to act on a range of individual posts and individual threads. These include the abilities to "merge posts, move posts, split topic and set invisible/visible" (Mytton, 2005, p. 123). Moderators can use these tools to channel the discussion in different directions. If they think a discussion is getting off topic, they can split it into multiple discussion threads; similarly, they can merge and move posts at their whim. The ability to make posts visible and invisible is intriguing. Instead of just deleting posts, administrators and moderators can hide them from particular kinds of users. As a result, there is a record of what was said, but it's just not visible to everybody. In short, moderators have been equipped with a significant number of tools to be able to tweak and change the order and structure of active and inactive discussions.

It's not just the content of the posts that they can control, however. Moderators in this particular system can also "unsubscribe all members from e-mail notifications for this topic" (Mytton, 2005, p. 124). Where the other options for inline moderation involve changing the content of the discussion, this method enables the moderator to change how users become aware of posts. In many cases, users posting in a discussion thread are automatically subscribed to emails notifying them of new comments in the discussion thread. As a practical matter, turning off email notifications for a particular thread is a potent way to shut down a conversation. Users can keep posting there, but none of the participants in the discussion will be notified that there are new comments. Now, any of the users can come back to see that there are new comments and ongoing discussion, but for those users who rely on emails to notify them of activity in their discussions, it will seem like the discussion simply ended. There is no visible indicator to anyone other than the moderator that this feature has been turned on in a given thread.

Think about the difficulties this imposes on making sense of a discussion thread. If halfway through a moderator grew tired of seeing a particular discussion float to the top of the list, he or she could quietly turn off email notifications for the thread. Users who manually visit and login to the site to see if there is new content will find the discussion and continue to participate in it. However, to all users who participate in the discussion based on email notifications, it will appear as if the entire discussion has ended. This illustrates a broader point about how moderation works in online community systems. In many cases, the interface to a community and discussion on the website is just one of a series of modes through which users interact with the community. In this case, users email applications become an interface to the content of the community. In other cases, users subscribe to updates for online community through an RSS reader, and still others might subscribe to a Google search alert for particular terms that Google finds in the text of a discussion thread, which can bring them in to participate in a particular discussion. The result of all this is that making inferences about the text of online communities based on an absence—for example, noting that a particular user who normally responds to this kind of issue has not responded to it—is incredibly difficult to do. There is a range of techniques, including turning off email notifications, that make it easy for users to simply be unaware that a conversation continued beyond their own comments.

Mass moderation tools. The inline moderation tools are relatively straightforward. The moderator has the ability to execute a range of additional actions on discussion topics and posts. In these situations, the moderator is interacting with basically the same interface that the rest of the users and visitors to the site see. Beyond this, many of these systems come with the ability to manipulate the content of discussion forums en masse.

Invision Power Board 2 comes with a tool called "prune/mass move" that allows admins and moderators to "move or prune many topics according to certain criteria" (Mytton, 2005, p. 126). I'll delve a bit into the criteria in a moment, but first, the term "pruning" comes up in many of the books on online communities, and it's worth considering the implications of it. Let there be no confusion: pruning discussion threads is tantamount to deleting them. With that said, the notion of pruning comes with the idea that much like a gardener prunes away parts of a bush or a tree, a community manager/administrator must prune away some of the discussion to make room for new growth. While there is often a persistent belief that the archive of the aggregate discussions is valuable, it is important to recognize that there is also the competing view that there is value in pruning away some of the discussions over time to keep an online community healthy.

The criteria for pruning and mass moving discussion topics offer further insight into the logic of the discussion board software. As Mytton explains, mass moving and pruning "can be done based on: topics with no new posts over a specified period of time, topic types (open, locked, moved topic links, or any topics), topics with less than a specified number of replies, topics started by a certain member" (Mytton, 2005, p. 127). The criteria-driven approach to mass moving and pruning discussions enables an administrator or moderator to leverage the data fields in the underlying database to identify discussion without even reviewing or rereading them. Administrators or moderators can erase every discussion begun by a particular user; they can delete all of the topics based on any number of different parameters that characterize those topics. In short, the moderator or administrator is empowered to make sweeping changes to the discussion content of the community. In contrast to the inline tools, these en masse tools enact a very different kind of agency on behalf of the administrator/moderator. In the inline moderation techniques, the moderator decides to intervene in a particular situation and uses the moderation tools to do so. In the en masse cases, it's likely that the individual moderator/ administrator will not really understand the ramifications of what he or she is doing to each of the individual threads that is affected. As a result, if one wanted to make inferences based on the fact that a particular topic has been removed or deleted from an online community, it is difficult to ascertain who did this and to what end. Where one might imagine that some moderator was attempting to suppress or remove a particular set of opinions or perspectives, it's just as likely that the deleted topic was either part of a continual plan for pruning content from the site, or that it was deleted as a kind of collateral damage resulting from some attempt to tidy up the forums en masse.

IP member tools. In a section titled "IP Member Tools," Mytton explains how administrators can view a considerable amount of information about site members' IP addresses. I'll let Mytton explain how IP addresses work: "Every machine and server on the Internet has its own unique IP address, which identifies it for the duration of its Internet session. This can be used to track down the ISP (Internet Service Provider) that the member is using, and its country. It can also be used by the ISP to track down an individual customer" (Mytton, 2005, p. 106). An Internet Protocol address is a prerequisite for interacting on the web. While there are ways to mask or alter one's IP address, it generally does provide relevant information about each user visiting a site. Invision Power Board provides a set of tools for administrators and moderators to use to record and use that information to moderate and ban users. I'll quote it at length to give a sense of exactly how this works in the system.

> IPB contains tools to allow you to find out which IP addresses your users have been using and to look up information about that IP. You can then use the IP to place a ban if it belongs to a troublemaker, or use it for troubleshooting purposes. Clicking on IP member tools will show you a screen with two boxes. The first allows you to see all the IP addresses that a specific member has used. Typing their name into the box will list all the IP addresses. You can then see how many times each IP has been used, the last time it was used, how many users have been registered with the same IP and a link to find out more. (Mytton, 2005, p. 106)

By logging and organizing information about every IP address that a registered user has used to log into the system, the Invision Power Board system, as well as the other systems built on the same principles, is able to make it easy to enact control through information recorded in the IP address. While anyone running a system can keep HTTP logs that will maintain this information, the design of these systems is set up in such a way as to make it as easy as possible to operationalize that basic information into information that can be used to try to ban users. This approach to using IP addresses comes up extensively in Patrick O'Keefe's book on managing and running online forums, which I discuss next.

Dealing with problem users, bans, and chaos. Patrick O'Keefe's 2008 book *Managing Online Forums: Everything You Need to Know to Create and Run Successful Community Discussion Boards* offers readers a wide range of suggestions for doing exactly what the title of the book suggests. The book is the result of O'Keefe's years of experience in running, managing, and administrating an array of online forums, including everything from SportsForums.net, KarateForums.com, phpBBHacks.com, CommunityAdmins.com, and Photo-shopForums.com. O'Keefe's book is directly focused on running, administrating, and managing web forums. He primarily discusses the functionality of two of the most widely used web PHP and MySQL-based platforms, vBulletin and phpBB.

While there is a considerable amount of content in the book that is potentially of interest, one section of the book provides detailed discussion of problem users and the kinds of tactics and strategies that one might invoke to curb and control their behaviors. The chapter "Banning Users and Dealing with Chaos" describes a range of problem users and how to deal with them. These problem users include everything from "adverquestions," in which new users show up and offer thinly veiled marketing messages, to "content thieves," who repost forum content elsewhere, to users like the "reply-to-every-post-guy," the "freedom of speech guy," who insists that "freedom of speech entitles people to say whatever they wish, whenever they wish, wherever they wish," which O'Keefe explains remains "one of the most common misconceptions and problem issues for community administrators." Other types of problem users described include the "I'm Creating My Own YourSite.com" user, and the most intense "Hate Him, My Minions! Hate Him!" in which someone who runs a competing online community site "becomes jealous of you and abuses his position to manipulate his user base" and sends them all to attack your site (O'Keefe, 2008, pp. 185–199). O'Keefe goes on to suggest exactly how to structure and manage forums to deal with these problem users.

Curbing abuse: Report buttons and automating face-work. After providing a range of suggestions for how to handle particularly difficult user situations, O'Keefe suggests that a post-reporting system is a great way to curb abuse—in this case, having your forums include a "report post" button next to each post that will add the post to a queue for the administrator or other site manager to review. Beyond including the button, he advises admins to encourage members to use the report button and to make sure that moderators use their judgment in deciding when to remove posts for violating the guidelines for the discussion board. The concept of creating and posting these kinds of guidelines gets its own chapter in many of these books. There is a technical system and an emergent social system in place here. The first feature includes the structural components, the report post button, and the queue of reported posts; the second requires getting participants in the discussion boards to use the report post button and finding and recruiting moderators who will read the queue and use their judgment to decide what is and is not a violation. This mixture of a technical system and social norms in effect implements a particular set of ideas about governance. Anyone can report anyone, and moderators judge whether the reported activity is in conflict with the site's rules or norms. Already significant to the study of online discourse is the fact that moderators delete and prune discussion on the site. Discussion threads are not direct transcripts of conversation—they can and often do change over time, particularly if they relate to hot-button topics.

O'Keefe then explains how "helpful notices" can affect posting. For example, on his phpBBHacks.com support forum site, whenever users start new posts, they are prompted with a prominent notice in red text that urges them to make sure that they are posting in the right section of the forums. As another example of a helpful notice, O'Keefe suggests the value of prompting users who respond to discussion threads that are older than a specified age that—again in red text—"The thread is X months old and that he might want to consider creating a new thread instead" (O'Keefe, 2008, pp. 204–205). Where the report post button functions to police posts in the discussion by helping identify inappropriate posts for removal, O'Keefe's helpful notices act to pre-police posts. The goal here is to influence posters at the moment they are about to post by giving them a particular bit of just-in-time guidance. In the case of the latter suggestion, this guidance can be programmed to appear only in particular kinds of discussions. Administrators are consciously deploying these cues to shape discussion. Thus, those studying communication in these kinds of sites need to think about how things such as the post box, or the posting page itself, may include this kind of just-in-time information in an effort to steer conversation in a particular direction.

Word censors. Under the heading "Innovative Tools," O'Keefe goes on to give two examples of ways in which a developer friend helped him by creating what he refers to as "hacks" for web forums that he manages. (O'Keefe uses the term "hack" to refer to extensions or plug-ins for software such as phpBB.) These hacks are interesting on a few levels—principally, the way in which they illuminate what he refers to as "automation" can directly affect the nature of online discourse. The hacks also offer a way to understand how guidelines, norms, and rules of a community site can be enacted as procedural or algorithmic rules. These scripts and hacks effectively become actors in the communicative discourse of the forum. They lie in wait and pop out at prescribed moments in discursive interaction to mediate and perturb the order of the communicative act. In this sense, the web forum and its hacks are something akin to robots participating in and altering the kinds of discursive interactions psychologist Irving Goffman described as face-work (1967). For example, consider how O'Keefe describes a particular problem on his phpBBHacks.com site.

> We had used the word censor to block out inappropriate language but I was thinking about that system one day and it dawned on me: What if those posts were stopped when the user tried to post them? And what if the community software explained why and even highlighted the sections of the posts where the violation(s) occurred, allowing users to make adjustments without losing their post? (O'Keefe, 2008, p. 206)

The "word censor" he describes is functionality that draws from a list of inappropriate words to block them when they are displayed on the site. It's worth noting that similar discussions of word censors show up in many of the books, in the context of their relative merits for controlling user behavior. Instead of obscuring censored terms, O'Keefe wanted his site to automatically reject posts with censored words. He then wanted to provide in-context information about what terms had triggered the censor. As a result, all communication on the site first involves a brief inquest from the site's censor. A post will either pass or fail, and if it fails, the user can make changes before the post is ever recorded. Where the previous censoring tool wouldn't display words on the censor list, this new plug-in won't let even traces of them remain. O'Keefe goes on to provide us with the text prompt he gives users who trigger the word censor.

> Your post has triggered our word censor feature. The portions of the post that triggered the censor are highlighted in the preview below. Please adjust it and attempt to post again. Please note that abbreviating the term/string that was censored or

circumventing our word censor feature in any way constitutes a violation of our user guidelines, and your post will be removed. (O'Keefe, 2008, p. 207)

When users are told what is being censored, they can easily work around it. Instead of writing "ass" you write "a$$," and you have tricked the word censor. This note is included with the reference to community guidelines in order to wrap a normative layer around the word censor functionality. Earlier in his book, when describing the concept of basic, built-in word censoring functionality, O'Keefe suggested the importance of this norming layer: "Don't forget, you can never censor every vulgar term. Don't even try. It's not possible." He explains: "People will use words or come up with new ones that you didn't or cannot censor and you will have to remove their posts." The lesson is: "do not institute an 'if it's not censored, it's OK' type of guideline" (O'Keefe, 2008, p. 25). O'Keefe wants to use the word censor to help automate part of the rules and norms of the site, but for him it is critical that the automated functionality not become the rules and norms of the site. His hack exposes the logic of the censor to users, inviting them to revise their comments, but at the same time it explains to them exactly how they could circumvent the rules. For O'Keefe, this is where the social contract of the site's guidelines becomes critical. The guidelines ensure that users don't game the censor, and if they do, those users invite a harsher reaction.

This brings up some significant considerations for studying discourse in these online communities. For O'Keefe, the benefit of this approach is clear: "This saves us time and it saves the member time—his post doesn't have to be removed and we don't have to document the violation and contact him, because the violation is never made" (2008, p. 207). There is a benefit both to him and to the community member: they both save time and avoid an altercation. With that said, there is no record of what happened. The user attempted to say one thing; the system politely asked them not to say it; the user self-censored based on that feedback; and all that remains is the result of this back and forth. In short, when this kind of functionality is enabled on a site, we are studying something that isn't so much a transcript of what was intended but instead a transcript of a conversation that was pre-censored at the point of origin.

The difficulties encountered in banning users. From here, O'Keefe describes several approaches to banning troublesome users. Each of the user guides for community discussion systems describes the methods it provides for banning users, and each mentions that it's generally a problematic issue. Any banned user can ordinarily just create a new account. As O'Keefe explains, "Idiots and bad people exist and you'll be dealing with them." For him, banning a user is something that the user brings on himself. "As an administrator, you are simply reacting to what a member does" (O'Keefe, 2008, p. 207). Most of what he says about banning is what one might expect: he identifies particularly egregious individual situations that might result in the need to ban a user and discusses minor ways in which a user might repeatedly violate the community guidelines in such a manner that he or she should ultimately be banned.

O'Keefe's extended discussion of methods for banning users and his reflections on those methods are critical for understanding how control is achieved in these web forums—specifically, how administrators shape online discourse from the technical level. O'Keefe first describes banning usernames. Banning a username keeps a particular user from posting to the forum. Here site administrators wield considerable control and power. They can turn on and off a given user's ability to discuss. However, usernames are relatively weak ways to control or exert power over the people who use those names. As O'Keefe explains: "The member may just sign up again, but the username is her identity on the site and should, as such, be the first thing you ban." There is nothing stopping this person from setting up another account and starting to post again. Thus, banning a username is not so much an exertion of technical control; people can sign up for accounts and start posting again. It is primarily about normative control—namely, publicly shaming someone and blocking them from participating in the community under a particular username and the identity that username represents.

O'Keefe then moves on to discuss another kind of control, that of banning IP addresses. He explains: "Your community software should allow you to check what IPs a user has posted from." If it turns out that a user "made all of the posts from one address," banning IP "may actually work." Even if they made most of the "posts from IPs that are similar except for the last few digits you can block an IP range and that also might work." Having noted this, he recognizes the substantial risks. "Besides not always being effective, it sometimes prevents other users on the same Internet service provider (ISP) from reaching your community" (O'Keefe, 2008, p. 212). While administrators exercise considerable control over their communities, the tools at hand—blocking usernames and IP addresses—are both relatively blunt instruments. An antagonistic user easily overcomes these methods. While banning an account is a trivial task, it doesn't actually stop someone from simply creating new accounts if they so desire. Similarly, one can get around an IP ban by using a proxy server to connect to the site from a different IP address.

Attempts to ban users based on information in the underlying communication protocols are not new to IP addresses. As mentioned earlier, in 1988 Bowen and Peyton noted that some BBS sysops set a configuration option to "Deny access to callers who use 300 baud" because they felt individuals with less expensive modems were more likely to be juvenile (Bowen & Peyton, 1988, pp. 74–75). Similarly, in 1998, under the heading "The User Authentication Quandary," Cliff Figallo described how some online communities were trying to use email accounts with particular Internet service providers as a way to block undesirable users. He began by noting the problem of using email addresses to authenticate users: "The user might have more than one email address or might give his friend's address" (Figallo, 1998, p. 102). Beyond that, there was the problem of "the increasing availability of free email services and addresses through sites that allow users to create them at will." In short, relying on email addresses as the unique identifiers for users was becoming messy, since a user could just go and create another email address and come back and harass you—to which Figallo responded: "This is resulting in some community sites refusing authorization for accounts showing email addresses not tied directly to an ISP." So only email addresses such as @comcast.com or @cox.com, addresses that can be traced back to a particular user, were being authorized. At the time, this was already falling apart as a mode for uniquely identifying users, as Figallo went on to note: "Unfortunately, the largest source of email addresses in the world, America Online, allows its users to create and change user names at will, and refusing anyuser@aol.com takes a big bite out of the potential pool of registrants" (Figallo, 1998, p. 102). The longstanding interest in creating ways to selectively ban potential users based on these kinds of procedural methods continues to prove elusive and failure prone. The structures of the underlying communications protocols are not particularly well suited to the kinds of control that site administrators would like to exert.

Creative approaches to banning users. In a section titled "Get Creative," O'Keefe explains a series of ways to thwart attempts from bad actors to access and interact with an online community. Creativity, in this case, tends to mean more levels of obfuscation and manipulation. First off, this includes making it look like your site is down. "You could make it so that a 404 (not found) page displays when a specific IP visits your community." He explains how you can configure a .htaccess file to display a 403 (forbidden) page to any user from a particular IP range. Here O'Keefe is using Apache and the HTTP protocol to shut down participation. But beyond the protocol, he suggests adding an additional layer of deception. He explains: "You can customize your 403 page to look like a 404 page, which will give the impression that the site is down" (O'Keefe, 2008, p. 214). O'Keefe is not simply suggesting that one should use the HTTP protocol to block access; rather, he wants you to take the additional step of misrepresenting what you are doing and making it seem like the site is down.

In case these ideas sound particularly extreme, so much so that you might think they are idiosyncratic to the author, O'Keefe goes on to explain some easy ways to "simulate downtime." These include the "Miserable Users" hack for vBulletin and "Troll" for phpBB, both of which "combine downtime, slowness, general confusion, and the site actually working." The hope of the "Troll" and "Miserable Users" hacks, like O'Keefe's suggestions for the 403 .htaccess hack, is that they will "frustrate these troublemakers and drive them away." Not only are O'Keefe's ideas about how to deal with troublesome users more widespread, but there was actually enough demand for such functionality that similar plug-ins were created for the two most popular discussion board software platforms.

The "global ignore" or "hall of mirrors." Instead of making it look like the site itself is down, O'Keefe offers a related approach for shutting particular people out of the conversation. "Sometimes referred to as global ignore, you can incorporate a function that lets the banned user log in but then makes the user go unseen to all users of your community." Such users think they are participating in the community, but in fact they are not. "He can still make his posts, but only he (and maybe you and your staff) can see the posts—no one else. Basically, in his eyes, the site works as is intended. He will, hopefully, just think that everyone is ignoring him and go away" (O'Keefe, 2008, p. 214).

The globally ignored user has been muted, a rather deceptive practice. It is hard to conjure up a comparative situation in other modes of communication. If you mute someone on a conference call so that only they can hear themselves, it quickly becomes clear that no one is pausing, waiting to talk, or doing any of the other things we do when we are engaged in communication. Because the globally ignored individual continues to see himself occupying the same space in the threaded discussion, it would likely take more time for him to realize what is going on in communication. This technique is used in other web applications as well. When offering advice to anyone designing social web applications in *Designing Social Interfaces*, Crumlish and Malone use a similar tactic described as the "Hall of Mirrors," explaining that you can put problem users "in a 'Hall of Mirrors' in which only they (and perhaps others who have been banned to the Hall of Mirrors) can see their posts." In this setting, "They will wonder why no one is falling for their tricks anymore" (Crumlish & Malone, 2009, p. 391). The global ignore or hall of mirrors illustrates the complex mixture of control and lack of control that administrators and moderators have. It works only as long as the user is logged in with that particular account. Seeing that they have been muted in this fashion could be as easy as logging out.

These deceptive practices illustrate a sophisticated mixture of extreme control and an extreme lack of control. The range of methods and approaches that an administrator can employ are part of a complicated social dance, and all of the technical approaches to banning and keeping users away come with significant limitations. An individual can simply sign up for a new account, or change his or her IP address. At the same time, it is clear from these extreme examples just how much power administrators wield in shaping and manipulating the experience of online discourse. The lessons for those interested in studying discourse and conversation in online communities should be the need for understanding the context of communication. There are clear obfuscatory practices that admins have at their disposal, and there is every reason to believe that particular discussion participants deemed to be bad actors are being silenced in any number of online communities. This is not to suggest that there aren't ways of finding evidence and information about this silencing. In many cases it would be valuable for a researcher to spend some time thinking about the ways users can be shut out of a particular online discussion, and where the researcher might find evidence either in the particular community or on other sites that explains how and why particular kinds of users have been silenced.

Given the problems that come with attempting to ban users, some designers of online communities have developed methods for publicly shaming users. In *Designing Social Interfaces*, Crumlish and Malone (2009) suggest "utilizing the technique of disemvoweling to censor unwanted comments or spam without having to actually delete the posts or comments" (p. 280). The idea of disemvoweling, which can be traced back to usage on Usenet in 1990, generally involves procedurally removing all of the vowels from a user's comment or post while leaving all the consonants. The technique, now used on a range of blogs like Gawker and Boing Boing, is intended to leave the user's comment somewhat legible, but also publicly shame the user in a way that deleting the comment wouldn't. As Crumlish and Malone suggest, by using the practice in public, the site sends a message that this behavior is "unacceptable"; further, it makes that particular "bad apple look stupid" (Crumlish & Malone, 2009, p. 280).

Empowering users to moderate collaboratively. Increasingly, the designs of these systems are adding particular functionality. One is the "report abuse" button mentioned previously that turns some of this moderation control over to the registered users of the site. As the idea of "social" comes to replace the idea of online community, we increasingly find these kinds of collaborative moderation systems in which users thumbs-up, thumbs-down, or give star ranking to other user comments, posts, or contributions, which in turn control what is and isn't visible in community discourse.

As Tharon Howard explains in the 2010 book *Design to Thrive: Creating Social Networks and Online Communities That Last*, you want to "allow users to read a posting by another member of the forum and to assign a score to the value of the contribution.... The scores of individual members can then be averaged over time and equations can be developed to allow those members of the community who have a high number of postings that have received a high rank to float to the top" (Howard, 2010, p. 66). Moderating content is distributed through the votes of logged-in members. Instead of being part of an ongoing threaded discussion, each comment belongs to a set of sortable, stand-alone bits of information, to be organized according to its popularity. Crumlish and Malone describe similar systems as "vote to promote" (2009, p. 266) and "systems for collective choice" (Crumlish & Malone, 2009, p. 267). In *Design to Thrive*, Howard goes on to explain that this approach has

> two positive effects in terms of remuneration for the community. First it gives members something to shoot for (as in a video game) and it has a certain entertainment value that many people enjoy. Second, reputation and effectiveness scoring has the effect of shutting down and discouraging inappropriate postings and ineffectual messaging techniques. It discourages people from participating in flame wars, name calling and *ad hominem* attacks, which are going to lead to lower average scores over time. (Howard, 2010, p. 67)

For Howard, this kind of thumbs-up/thumbs-down functionality both incentivizes the kinds of behavior he wants to see on his site and disincentivizes the behaviors he doesn't want to see. In this sense, it is a moderation technique that helps bring to the surface the behaviors he wants to see modeled and replicated by users. In contrast to previous approaches to deleting or removing discussion content, these techniques simply shuffle it out of site.

The result is that it becomes increasingly difficult to figure out exactly what a particular user would have seen at a given time. Whereas you can read backward in a discussion forum and generally feel confident that you are looking through an ongoing discussion, when comments and user responses become something that is shuffled and sorted based on things such as votes, it becomes increasingly difficult to piece together what a given user may have seen in a given situation.

Moderate, ban, filter. This survey of the tools and techniques that allow site moderators and administrators to moderate and filter content and ban users suggests the staggering complexity involved in making sense of the records that online communities leave behind. Likewise, the development and nature of these tools illustrates the complex struggle that emerges between attempts to moderate and control users. Users can continue to stir up trouble in that it is very difficult to find a technical way to identify them and keep them banned if they continue to cause trouble. They can always simply start up another account. Those points aside, this entire section further underscores the logic of control and management at the heart of the idea of online communities. These are places created and managed toward particular ends by the administrators and moderators who set up and maintain them. Further, much of their power and tools are not obviously displayed. Because it is so easy for users to come back and cause trouble, many of the tools—such as the global ignore or the ability to just stop sending out email notifications—are rather coy attempts to use a kind of soft power to manipulate and push online community discourse in the direction that the site management wants it to go in.

Reputation systems. At this point, much of our experience on the web is structured and created based on sets of underlying reputation systems. In the 2010 book *Building Web Reputation Systems: Ratings, Reviews & Karma to Keep Your Community Healthy*, Randall Farmer and Bryce Glass explain that designing these systems is "one part each engineering, social science, psychology, and economics" (Farmer & Glass, 2010, p. x). That is, the design models for ranking users and users' contributions to sites are as much a form of social engineering as they are technical. From Farmer and Glass's perspective, explicit reputation models and systems are "critical for capturing value on the web, where everything and everybody is reduced to a set of digital identifiers and database records" (Farmer & Glass, 2010, p. 20). A far cry from the utopian visions of 3D avatars finding spiritual enlightenment on the web in the 1990s, the systems that undergird the web today are increasingly based on the assumption that everything and everybody is reduced to records in a database in the system.

This section briefly tours the origins of some of these reputation systems, looking first at their development in BBSes and in online community discussion board platforms. From there, the section delves into recent work on reputation systems in social web applications to document the increasingly sophisticated developments of social theories of user action and agency that inform the design of web applications. This analysis examines the way in which the idea of distributed and collective intelligence is working its way into the design of online community. Increasingly, the view of an online community is itself organized and created as the result of analysis of the reactions and actions of other community members. The entireties of these systems treat user action as inputs that inform and structure the resulting outputs. As these systems become more and more sophisticated, they are developing a deeper respect and acknowledgment of the reactive nature of how their users engage with these systems. Interestingly, the result is that more and more of the nature of these reputation systems is actually being obfuscated from end users, instead acting to control the internal functionality of systems.

The origins of reputation in the database. While the term "reputation system" for online communities didn't come into use until the late 1990s and early 2000s, the fundamental principles that make it possible were there even in BBSes. For example, the following explanation of how a BBS functions from Mark Chambers's 1994 book *Running a Perfect BBS* illustrates the emerging logic of the reputation system.

> Each time an existing user logs onto your BBS, the bulletin board program loads part of the corresponding userlog entry into memory and keeps a record of how many files the caller receives from the system, how long the user remains online, and the number of messages the caller writes. This statistical data is often used by the bulletin board program or external utility programs to create reports, flag those who are not participating (by sending files to your system and entering messages), and award extra time and services to those members of your system who do more than their share. (p. 23)

From the bottom up, online communities are powered by database logic. Users' actions are logged and loaded and can easily be compared and contrasted to those of other users. In Chambers's explanation of this process we can see the beginnings of the logic of the reputation system. As the sysop, you can use the data recorded in the logs as the basis for punishing and rewarding users based on the logged records of their activity. All you need to do is identify the behaviors that you want to encourage, and you can provide those users with additional privileges.

Institutionalized gossip. The first of the online community books to explicitly talk about reputation systems was Amy Jo Kim's 2000 book *Community Building on the Web: Secret Strategies for Successful Online Communities.* Kim's discussion of reputation systems comes under the heading "Institutionalized Gossip" (p. 109). As she explains, "In both offline and online communities, people develop reputations primarily through word of mouth." That word of mouth is the gossip part. Kim goes on to explain how it becomes institutionalized: "You'll need a more explicit way to track and display reputation in communities where actions speak louder than words, where those actions are measurable, and where they're meaningful for decision-making" (Kim, 2000, p. 109). Identifying modes to measure actions that are meaningful to decision making is the crux of the reputation system for Kim. She uses eBay as a model reputation system.

> In a commerce-based community, knowing someone's trading history is an important criterion for a member contemplating a financial transaction. Were the goods delivered as advertised, and in a timely manner? Did the payer's check clear? On eBay, buyers and sellers start off with neutral, blank-slate reputations, which change as other members rate their interactions.... each person's eBay reputation is a cumulative summary of the positive, negative, or neutral marks that others assign to them. (Kim, 2000, p. 109)

Here, the cumulative ratings of transactions from one particular eBay user become the basis of that user's aggregate reputation score. It is helpful information for making a decision about whether this particular person is trustworthy. Importantly, Kim goes on to note: "In a conversation-oriented web community, implementing an explicit reputation system is tricky, because reputation is highly subjective in social interactions." While it's easy to see how rating the transactions on eBay is helpful in measuring the reputations in the context of selling things, the subjective nature of evaluating other social interactions leads Kim to suggest that it's better to avoid building explicit reputation systems for these kinds of activities. She goes on to explain: "Unless there are specific, relevant actions to measure, a reputation system is usually not called for—it's better to rely on word of mouth" (Kim, 2000, p. 109). Since the time that Kim wrote this, however, there have been many increasingly sophisticated attempts to measure and monitor users' reputations and to use those measurements to prompt users to take the kinds of actions that a site owner wants them to take.

Ranks in phpBB, vBulletin, and Invision Power Board. As online discussion board platforms such as phpBB, vBulletin, and Invision Power Board rolled out in the early 2000s, they came with some relatively simple reputation systems built into their core design. Members of online communities running on these platforms are likely aware of the way the systems calculate and display ranks.

Under the heading "Manage Ranks," Mytton (2005) explains the functionality and goals of these rank systems in the user guide for Invision Power Board 2.

> Ranks are used to encourage members to post. As their post count increases, they will rise up the ranks, and gain status within your community. Used well, they encourage more posting and therefore greater activity. You can define your own member ranks from the Manage Ranks option. From a fresh installation, you will see three already defined: Newbie (0 posts) Member (10 posts) Advanced Member (30 posts). Once a member reaches the Min Posts value, their rank will be automatically changed. "Pips" (or a custom image you can create) will appear beneath the member's name in their profile and in their posts at this point. (Mytton, 2005, p. 96)

The logic of this system is simple enough. You want your users to post more, so you track their post count and give them different markers of status based on those ranks. What Mytton does not mention is that these attempts to encourage members to post often result in people gaming the system. While this system can encourage users to post, it doesn't say anything about the quality of these posts. For example, in one phpBB forum I participated in, there was a whole off-topic section of the forums where some users would regularly create discussion threads simply to post hundreds of messages to raise their post count to advance to higher ranks. As Amy Jo Kim has suggested, it is very difficult to implement systems that measure user action for these kinds of systems, as it's hard to automatically quantify exactly what is valuable about a user's response. With that noted, it remains curious that all of the major discussion board platforms developed at this time come with these kinds of ranks enabled out of the box. Despite the fact that these features were already known to be problematic, it was just assumed that this kind of functionality was required to be able to craft an online community system. As the design of online community systems has matured, one of the most extensively developed features is a far more nuanced approach to designing and modeling reputation systems.

Reputation at the heart of designing social. "What is reputation in an online community? In its broadest sense, reputation is information used to make a value judgment about an object or a person." So begins Randy Farmer and Bryce Glass's 2010 book *Building Web Reputation Systems: Ratings, Reviews & Karma to Keep Your Community Healthy* (p. x). In this 300-page book from O'Reilly Media, the authors provide a comprehensive approach to modeling, designing, and encouraging users to become inputs and outputs for the management of user-generated content.

The reputation statement. For Farmer and Glass (2010), the crux of reputation systems is the development of reputation statements. For these authors, reputation at the most base level is summed up as "a source makes a claim about a target." This could be something explicit, like "Bill says Harvard is expensive" or implicit, like "Wendy chose Harvard over Yale" (Farmer & Glass, 2010, p. 7). In both cases, one can make inferences about the target based on this relationship. For these authors, the process of designing these systems becomes about defining the targets of reputation, identifying sources to develop opinions, and codifying the claims that sources can make about your targets. For example, the user Stan has given a four-star rating to the movie *Aliens*, which can then be used to aggregate a collective rating for the movie. This example illustrates a core point of their approach: reputation is their generic term; for them reputation is a property of objects as well as individuals. So, in this case, Stan is contributing to the reputation of the movie in this particular site.

Much of the book is about when you should and shouldn't show users your reputation system calculations. There is considerable acknowledgment of the fact that significant reactivity comes from users seeing this kind of information and changing their behaviors. In general, they suggest shying away from showing reputation scores for people and instead focusing on reputation scores for objects (particular reviews, comments, videos, etc.).

To provide a sense of how complex the systems are that Farmer and Glass are explaining and suggesting that developers consider creating, consider their explanation of Flickr's "interestingness" system. Briefly, the photo-sharing site Flickr defaults to ordering images on its site according to their "interestingness." The interestingness of a photograph is the result of a wide range of actions any number of users take to interact with the photo. Combined with information about each user who views the photo, the result is a score for each particular image that informs an overall rating of interestingness of the photographer who took and shared the photo. All of the activities users engage in with each photo are weighted through a series of filters—in this case, the stronger a relation-ship that viewers of a photo have to the photographer in terms of connections within the social network component of Flickr, the more weight ascribed to their actions on the photograph. Through this range of inputs and weights a composite score is calculated daily for all of the photographs on the site, and the 500 most interesting are shown on the "explore" page of the Flickr website. The actual scores are not made public, nor is the exact process that determines how the scores are calculated, and the order of the 500 photos is actually randomized to make it less clear exactly how this calculation was made. As Farmer and Glass explain, "randomness makes it nearly impossible to reverse-engineer the specif-ics of the reputation model—there is simply too much noise in the system to be certain of the effects of smaller contributions to the score" (Farmer & Glass, 2010, p. 89). Given that people would try to game the system if they could figure out the exact nuances of how it is weighted, the decision to obfuscate how exactly the scores are generated serves to help the designers of Flickr encourage the behavior they want to see in general, without giving enough information to make it easy for users to focus on how to game their system.

The example of the Flickr interestingness reputation system illustrates the increasingly sophisticated way in which user actions are becoming part of a network of input that creates and structures the outputs those users experi-ence in online communities.

Designing for reactivity. As already mentioned, one of the core principles of Farmer and Glass's approach to designing reputation systems for online communities is to consider the reactive nature of the systems that one imple-ments. For example, here is how they explain why they are generally opposed to the use of leaderboards for ranking users based on their contributions.

> The most insidious artifact of a leaderboard community may be that the very presence of a leaderboard changes the community dynamic and calls into question the motiva-tions for every action that users take. If that sounds a bit extreme, consider Twitter; friend counts and followers have become the coins of that realm. When you get a notification of a new follower, aren't you just a little more likely to believe that it's just someone fishing around for a reciprocal follow? (Farmer & Glass, 2010, p. 195)

In incentivizing users to take particular actions to make it onto a leader-board—to climb the ranks—they suggest that the designers of a site call into question the motivations behind the behaviors that they want users to engage in. The example from Twitter underscores this. By prominently placing the metrics of followers on the site, the site is explicitly encouraging people to treat those numbers as the metrics of value in the community, and it changes the nature of the way in which people interact. No longer are people just following folks because they are interested in hearing from them. Rather, they are trying to see if they can gain them as followers as well.

Ultimately, Farmer and Glass are rather wary of approaches to showing metrics of reputation in online communities. Instead, much of their advice focuses on how to create models that function to sort, organize, and filter user-generated contributions like comments, reviews, and such and to rank and weight users' evaluations of each other to decide what content is shown to what users in what contexts. It's intriguing that their work—and many of the similar kinds of ideas that come out of Crumlish and Malone's (2009) *Designing Social Interfaces*—largely focuses on attempting to obfuscate many of these metrics at the same time that educational technology enthusiasts have become particularly excited about notions of gamification, exemplified in a range of educational research projects. Such projects, funded by the MacArthur Foundation, are focused on digital badging and more or less bring over exactly the kinds of features and functionality that the designers of online community reputation systems think are not particularly useful—namely, incentivizing users with badges and maintaining leaderboards to rank users based on their activity.

Database record being. As the concept of the social web has eclipsed much of the notion of online communities, the reputation system has increasingly become the logic that defines and structures our experiences with each other over the web. As Farmer and Glass (2010) explain, measuring and creating loops of feedback between input metrics and output metrics are essential to structuring interaction on the web, "where everything and everybody is reduced to a set of digital identifiers and database records" (p. 20). From the point of view of the reputation system on a site like Flickr, users are the source of inputs that help to define the results of the system as viewed by other users. Interaction and exchange between people are increasingly not just computer mediated, but are rather actively structured, designed, and engineered. The kinds of face-work in social interaction that Erving Goffman (1967) described are becoming more and more enmeshed in a process where algorithms and models that govern what is seen, and when, are themselves actors in communication.

· 7 ·

STUDYING THE RECORDS OF
ONLINE COMMUNITIES

The 28 books analyzed in the study that this work is based on result from, and articulate, perspectives in the discourse of the design and implementation of the server-side software that enables online community. The tactics discussed by each of these books' authors suggests the goals and values of their work. Analysis of the books demonstrates the role software plays in online discourse. This analysis also documents design elements that researchers interested in studying the textual record of online discourse should attend to in research on online communities. In this chapter I briefly present a set of questions for researchers to ask of the records of any particular online community and offer some strategies for trying to mitigate some of the inherent problems involved with the interpretation of online community records.

Through a historical review of the rhetoric of online community, I've suggested that the dominant ideology of online community establishes websites that grant permission, to varying degrees, to different users toward the goal of conditioning those users to exhibit the kinds of behaviors that the owner of an online community wants them to exhibit. While there is a persistent mixture of desire on the part of users or participants in online communities to have a role and a voice in those communities, such desires stand in marked contrast to the explicit modes by which control is enacted in those online communities. By and large, online communities are governed by a logic of ownership, control, and limited permission, and as such it's important to approach the records of these communities with skepticism in terms of what they actually represent about the views and perspectives of members and participants.

Control, Empowerment, and Its Limits

Visual design, information architecture, text prompts, and reward systems are all designed with the intention of stimulating particular types of discussion between particular kinds of users. Much of this is evident to someone who is willing to read closely the resulting interfaces. The placement of a post button and the structure of a reward system both leave interpretable traces in the rendered pages on a user's screen. As is evident in O'Keefe's examples of extensive manipulation of problem users, administrators have considerable power to control what is passed to a user's web browser. At the same time, the problems with different techniques for banning users illustrate the fundamentally limited nature of those control techniques. The technical means of control are limited by the ability of users to change their usernames or IP addresses. The administrator's tools for banning a person are all tools for banning poor surrogates for actual people, usernames, and IP addresses. The forum administrator depends on the norms and rules established in the posted textual guidelines for the site to establish and retain control.

Analysis of these books offers a useful means to further triangulate how power and control work on the web. The tension between control and empowerment found in descriptions of the tactics used provides nuanced validation of the same tensions Alexander Galloway found in his 2006 examination of the specifications for TCP/IP and DNS in *Protocol: How Control Exists after Decentralization*. Wendy Chun's argument that the web was "sold as a tool of freedom" but can also be understood as a "dark machine of control" (Chun, 2005, p. 2) could very well be applied to this brief description of the history of online communities. With that said, the idea of virtual utopia giving way to dystopia oversells the experience of these communities in practice. One could alternatively suggest that an era of boosterism for the high-minded potential of these systems gave way to a much more mundane set of practical and more technical presentations of how particular systems work in practice.

Theories of Users as Generalized Others in Design

These 28 texts espouse theories of users. They each offer taxonomies of users. In this respect, the texts offer insight into the ways that administrators and developers think about their users. Exploring how these authors describe good and bad users—whether as engaged and on-topic participants and commentators or as the people who make them money—offers a point of entry into how the idea of these user types plays a role in the design and functionality of these systems. The various ideas of the good and bad user are sets of expectations through which functionality is described and explained.

These generalized others—these ideal types of users—play a key role in the design and configuration of online community software. The ideas of different kinds of users serve as warrants in the arguments that each of these authors presents for why one should design and implement software in a particular way. Studying these texts suggests models for how developers' ideas of particular theoretical good and bad users play a role in their design decisions, which in turn are manifest in the actual material affordances of the software itself.

These theories of users are committed to the page and disseminated as cultural scripts for other developers and administrators to look up and potentially integrate into their internal theories of users. Beyond being inscribed on pages, these ideas are also encoded into the procedural scripts in the server-side software we interact with as users. The ideas of users depicted in these texts become actors in our social interaction as they are operationalized into the design of the software itself.

Software as actant, actor, and mediator in face-work. Throughout the stories and advice offered in these texts, the functional and structural characteristics of these software systems can be thought of as interjecting themselves as actants, as mediators, or as procedural participants in the face-work that occurs in online communities. In the case of the excerpt from Derek Powazek (2002) about the post button, the designer has interceded to structure the visual layout and interaction of the site to act as a filter for particular kinds of interaction. While he is not blocking out particular kinds of users, he is explicitly suggesting that designers put together the infrastructure of online

communities in ways that promote particular kinds of values. In the case of
Anna Buss and Nancy Strauss (2009), we find suggestions from designers to
use a behaviorist theory of the self and society to underpin a system of rewards
and encouragements doled out by the online community software to reinforce
and reward particular behaviors that drive profit. From reading these texts we
cannot know how individual users perceive or are affected by these mediators
of interaction, but we do gain an understanding of how developers and administrators might be thinking about and designing for particular identified,
intended behaviors.

The software is an actor, one that carries traces of the platform, of the
plug-ins and hacks, of the administrator's decisions. The scripts that pop up
and give helpful advice, the plug-ins that make it look like the site is down,
the reputation system that tells me I have 200 points—all are simultaneously
manipulating me to behave in particular ways, enabling and empowering me
to engage in specific behaviors, and suggesting ways others should engage with
and treat me. These how-to texts offer an approach for identifying how developers and administrators design for particular intended behavior and embed
their values and ideas into the functionality of these systems. While designed
with a particular set of theories and values about communication and users,
the resulting software then acts with and upon users. Users interpret, work
around, and use the software to their ends, and their use and action is itself
interpreted and evaluated by the developers and admins. Developers, admins,
users, and software intermingle, continually configuring and reconfiguring
online discourse. As much as systems are configured to users, we users are
configured to our systems.

Configuring the Knowledge Base, Collective Intelligence, and the Mind.
When we Google for the answer to a question, when we consult restaurant
reviews on Yelp from our mobile devices, when we troubleshoot a problem
with the answers found in a thread in a discussion form, we are consulting an
extensive world knowledge base. Taken together, the series of technologies
that create our experience of the World Wide Web congeal into a massive
cognitive prosthetic device. This collective intelligence is enacted through
the configuration and administration of software described in this book.

Understanding how the ideology and values of the designs of these systems work is not simply an issue of understanding technology. These systems
are part of our collective cognitive infrastructure. We use them daily as part
of our process of thinking. The product of online communities—their discourse in text—functions as a prosthetic memory. The algorithms of search

and filters in these systems act as synapses in our collective prosthetic mind. While much of our cognition and thought is mediated and distributed, online community software systems have been explicitly designed and refined with the intention of using us as inputs for information and moderation. To this end, it is particularly important to recognize and understand the values and ideology that are written into the structure and function of these systems. It is critical to appreciate exactly how control is enacted in these systems. As we use and are used by these systems, the decisions involved in their administration, design, and implementation increasingly structure our cognition.

Applied Implications of Understanding Online Communities

Aside from these general conclusions about understanding the relationship between online communities and our general understanding of the web, I want to briefly offer a set of particular implications from this work for those interested in studying online communities, as well as for those interested in using or building online communities for educational purposes.

Implications for studying online communities. In establishing a research method for netnography—ethnographic research on online communities—Kozinets suggests that the wide availability of archival data in old discussion boards and listservs provides the equivalent of "every public conversation being recorded and made available as transcripts" (Kozinets, 2010, p. 68). While he recognizes that "the nature of the interaction is altered—both constrained and liberated—by the specific nature and rules of the technological medium in which it is carried" (Kozinets, 2010, p. 68), his idea that the textual records of content from online discussion boards and other mediated communications are the same as transcripts of public conversations is fundamentally problematic, given the results of this analysis.

The fact that there exists a set of tools available to those who run and manage online communities to manipulate, reorganize, and otherwise exercise control over online discourse means that any attempt to study the records of online communities must begin with the assumption that what one has access to is partial, fractured, and likely reflects the desires and whims of the person who continues to make them available.

Someone interested in studying a particular online community should start by understanding the underlying design and structure of the community.

The historical chapter of this book (chapter 5) offers insight into some of the ideas at play behind particular tools and their functionality, which can help someone interested in studying a particular community to contextualize his or her work in terms of the tools in the given system. For example, knowing that a tool such as Invision Power Board v2 by default creates roles of Newbie (0 posts), Member (10 posts), and Advanced Member would keep an analyst from inferring intent of the particular administrators of the community to these labels. Knowing that these are the default terms in the software and not the explicit choice of the individuals running a particular site might even, in context, contribute to understanding how little care the administrators of a site have taken to customize and configure it. Aside from specific examples of individual platforms documented in this book, the general approach to finding user guides and manuals for running the software of a particular online community can likely offer this kind of contextual information.

Returning to the idea of netnography, Kozinets suggests that when discussing the online records of online communities available in discussion boards and listservs, "Archival cultural data provide what amounts to a cultural baseline. Saved communal interactions provide the netnographer with a convenient bank of observational data that may stretch back for years" (Kozinets, 2010, p. 104). The notion of a transcript suggests that one has access to a full and complete record of discourse. As I've already suggested, given the results of this research, it is a fundamental mistake to label the records of online communities as recordings or transcripts. Intriguingly, Kozinets's suggestion to think of them as "archival data" offers a much more nuanced way of approaching them. Where the analogy to a transcript suggests a complete record of discourse, a robust sense of the idea of the archive suggests the partial nature and the important role that power plays in the constitution of that partial record.

If we move away from the popular notion of archives as any set of old material and shift toward understanding some of the research done on actual archives—places that collect, preserve, and provide access to historical materials—I believe we find a better place to think about studying the records of online communities. In *Silencing the Past: Power and the Production of History* (1995), Michel-Rolph Trouillot examines the historical record and historiography of the Haitian revolution. One of his principal contentions is that while it's common to think of archives as places that preserve the past, it is equally correct to think of them as institutions that produce silences of the past as well. Archives select, preserve, and provide access to materials along with their mandates, and in the case of the colonial archive that Trouillot considers, historians interested in understanding the experiences of those who were colonized need to read between the lines of whatever material remains in order to try to understand the perspective of the colonized.

A comparison between the colonial archives that Trouillot studies and the archived text in an online community might seem overblown, but while there is a completely different degree of control and power at play in the colonial archive and the records of some anime discussion forums, the same principles are at work. What persists is what those in power have a vested interest in making persist, or at least didn't care enough to erase. From this perspective, it is best for those interested in studying online communities to approach the extant material they can find with the same kind of skepticism that a historian would bring to studying records in a colonial archive. Furthermore, they must bring an awareness of and sensitivity to the kinds of controlling actions the technical infrastructure of the conversation and discussion that originally produced those records afforded/encouraged. It's important to ask: Whose voice is heard here? How do I know this is what it purports to be? What parts of this set of records are missing? Who constituted this collection of records, and for what purpose? Finally, where might I look in this data for perspectives and points of view that differ from those who had the power to decide what is and isn't kept?

Implications for educators interested in using and building online communities. Many of the assumptions about community embedded in the design and rhetoric of online community are likely not very similar to the ideas and notions of community that educators bring to the table when they think about building online communities. It is all too easy to pick up and run with the tools and visions that Silicon Valley and venture capitalists have developed and embedded in the design of these systems. When we think about using these tools—for instance, when educators embrace badging systems based on web forum and online community systems that focus on providing users with digital badges to reward them for meeting particular criteria for participating in an online community—it is important to remember that these are by and large tools created to extract value from users. They are not designed with the goal of creating public good, even if they sometimes wear that language.

I would suggest that educators and individuals interested in the potential role that online communities can play in civil society revisit some of the initial visionary ideals of online community. In particular, be suspicious of anyone talking about online community without discussing the notion of governance. In keeping with that perspective, it is likely a good idea for educators to spend time with books like Jono Bacon's *The Art of Community* (2009), which focuses on the logic that supports and animates the open-source and free-culture movements.

As educators study and develop systems that give out badges, such as those funded through the MacArthur Foundation's Digital Media and Learning grants to create online badge systems, it is important to understand the ideas and ideology behind their design. Briefly, much of the motivation for and explanation of the value of digital badge initiatives is grounded in a desire to know how reputation systems work in particular online communities such as Stack Overflow and Slashdot (Carey, 2012). In this respect, the rather sophisticated approaches to thinking about the reactivity of users in different designs of reputation systems is likely to be useful to those endeavors. With that noted, I encourage educators and educational researchers interested in these topics to recognize that the design of these systems is grounded in a particular set of ideas and theories about human motivation, an area where educational and developmental psychologists can make a considerable contribution to this conversation.

Specific questions to ask of records of online communities. The broad trends in this research are useful in their own right. However, I think they are particularly useful for directly informing research practice for studying online communities. To make the results of this research more directly actionable for those interested in studying online communities, I have prepared the following list of questions to ask of online communities.

1. *Visual design and text*: Where is the call to action? For example, where is the post or comment button on the screen, and what might this suggest about the context a user may have before posting? How and where are members invited to post? What is the site's tagline, and what kinds of cues does the tagline suggest about who is and isn't invited to participate? In each of these cases, think about how the design, layout, and text invite particular kinds of participation from particular kinds of users while dissuading others from participating.

2. *Steps and processes*: Piece together what the sign-up and moderation processes are for the site. You can likely find information about this in posted policies and rules for the site. Think through how these processes might act as filters to sort out what kinds of people are able to say what kinds of things. For that mater, think about how the process might set the tone for how users should participate or behave.

3. *Reputation systems*: Is there an indication of the presence or absence of ranks and roles for users? The reputation system might suggest a particular perspective from the owners of the site on how to motivate and nudge members to post more.

4. *Design decisions and community platform defaults*: Before inferring what a particular feature in a site means, identify whether it is in fact just part of how a particular platform works. Is a feature of a site a design decision of the site owner/administrator, or is it a default or generic feature of the platform being used? Keep in mind that in many cases a particular kind of rank or reputation system might be there just because it's a default. With that said, realize that the decision to use a particular system (for example phpBB or Vanilla web forums) is likely to be something worth considering in terms of what it suggests about the site owner's values and perspective.

5. *Relationship to site owners*: Consider what kind of relation is likely to exist between the site owners and the site users. For example, if you were interested in the perspective of the men who participated in the L'Eggs pantyhose site's online forums, the fact that those men are directly at odds with the site owners' vision suggests that it could be difficult to find their voice in records of the community that the owners maintain. If you have reason to believe that the members you are interested in might be at odds with the goals of the site owners, you should be all the more suspicious and critical in approaching the records of the online community.

6. *Historicize interpretation and analysis*: What year(s) are these records from, and what ideas and values were at play during that particular historical period that might have implications for your interpretations? In this book I've suggested a set of historical periods that surface from reading these how-to guides. Re-read some of the information about a particular historical period and think about how that might change your approach and your interpretation.

Strategies to consider. Alongside those targeted sets of questions, I have also pulled together a more broad set of strategies. Working from the assumption that one shouldn't trust the records of an online community as a transcript of communication, each of these strategies could be used to help triangulate what actually occurred in a particular online community.

1. *Seek out third-party archived websites*: For example, see if the Internet Archive or other organizations that participate in the International Internet Preservation Coalition have copies of the community. Web archives provide snapshots of what an online community's pages looked like over time, so they are likely to reveal discrepancies or removal of parts of the records of a community.

2. *Look for traces of absences or silences in the records*: Take the approach that historians take to colonial archives. Look for evidence of absence, such as discussions that don't make sense because there are references to users' comments that are no longer there, or missing date ranges in parts of the discussion threads. Given that many of the tools for web forums and community platforms involve tools for mass moderation and removal of posts, it's likely that there will be entire user contributions removed or blacked out, or significant missing sections in many records of online communities.

3. *Seek out moderators and users for their perspectives*: Much of an online community is likely to be hidden, visible only to particular users with particular permissions. So if you can identify and locate someone with whom to discuss this, you have a good chance of identifying what absences and biases might be evident in the records you have to look at.

4. *Look for perspectives on the online community from other online communities' records*: Look for discussion of the community you are studying in the records of other online communities. Different communities are likely controlled by different sets of administrators and moderators, and you could very well surface opinions and perspectives on the community you are studying by seeing how it is discussed and described in the records of other online communities.

Conclusion: Further Disaggregating the Internet

One of the ideas governing the framework for this book was to take seriously the suggestion made more than 10 years ago by anthropologists Daniel Miller and Don Slater that social scientists studying community interaction on the web need to focus on "disaggregating the Internet, that instead of looking at a monolithic medium called 'the Internet,'" it is critical to focus on the range of practices, software and hardware technologies, and modes of representation and interaction that may or may not be interrelated by participants, machines, or programs (Miller & Slater, 2001, p. 14). This disaggregated vision of the Internet is the vision of online communities we find through the discourse evident in how-to books for administrators and managers of online communities.

One of my biggest takeaways from this research is the importance of zeroing in on the particular features of individual online communities and the interplay among the software, administrators, moderators, underlying infrastructure of the web, and members and participants that produce community on the web. Along with this set of relationships, the idea of the user in the mind of the site administrator and developer is critical. In this respect it's not simply the relationships among software, administrators, moderators, members and participants, but also the ideas that those actors have about each other that play a key role in shaping the kinds of computer-mediated facework that occurs in online community. I hope that social scientists interested in working with the records of online communities find the questions and perspectives I have developed here useful to their studies.

APPENDIX

Example Data Collection Sheet

Overall summary, notes, and commentary on the book (2–3 paragraphs).

1. **Author Bios:** How does the author present himself or herself? What is the author's expertise? What are the back-of-the-book blurbs that are supposed to convince me of the author's expertise?
2. **Introducing Online Community:** What is the point? Big-picture definitions of web communities, author's goals for the book. Who is identified as the audience?
3. **Technologies in the Book:** What technologies are discussed in the book?
4. **Visual Design and Information Architecture Stories:** Give a brief description with page numbers and pull quotes. After the quotes, offer a tentative theory of underlying values/ideology.
5. **Moderating Content and Users:** Identify discussion of banning users, filtering comments. In each case, note if there is or isn't deception. Look for explanation of ethics, that is, when is it OK and when is it wrong, the ethical dimension. Is there a section in this on explicit rules and policies for users? Is the freedom of speech story present? After collecting quotes

and making these notes, offer a tentative theory of underlying values/ideology.

6. **Reputation systems:** Is there discussion of giving users points or badges for participation? This would include everything from post count to post ranks to actual badging systems. Look for watchwords such as "social hierarchy." Also look for counterarguments about users "gaming" these systems and decreasing the quality of discussion. **After collecting quotes and taking notes offer a tentative theory of underlying values/ideology for each.**

7. **Theories of Users:** Document incidents in which kinds of users are described—troublesome users, helpful users, users who could become moderators, etc. Build out lists of kinds of users in each of the texts. After collecting quotes and taking notes look over the whole list. How is this similar to or different from the other books? What kinds of users show up in books that focus on which kinds of tactics and approaches?

REFERENCES

Akrich, M. (1992). The de-scription of technical objects. In W.E. Bijker & J. Law (Eds.), *Shaping technology/building society: Studies in sociotechnical change* (pp. 205–224). Cambridge, MA: The MIT Press.

Bacon, J. (2009). *The art of community: Building the new age of participation.* Newton, MA: O'Reilly Media.

Bell, D.J., Loader, B.D., Leace, N., & Schuler, D. (2004). *Cyberculture: The key concepts.* London: Routledge.

Black, R.W. (2005). Access and affiliation: The literacy and composition practices of English-language learners in an online fanfiction community. *Journal of Adolescent & Adult Literacy, 49*(2), 118–128. doi:10.1598/JAAL.49.2.4

Black, R.W. (2008). *Adolescents and online fan fiction.* New York: Peter Lang.

Bogost, I. (2011, May 3). *Persuasive games: Exploitationware.* Retrieved October 25, 2014, from http://www.gamasutra.com/view/feature/6366/persuasive_games_exploitationware.php

Bowen, C., & Peyton, D. (1988). *The complete electronic bulletin board starter kit.* New York: Bantam Books.

Bryant, A. (1995). *Growing and maintaining a successful BBS: The sysop's handbook.* Reading, MA: Addison-Wesley.

Buss, A., & Strauss, N. (2009). *Online communities handbook: Building your business and brand on the web* (1st ed.). Indianapolis, IN: New Riders Press.

Carey, K. (2012, November 2). Show me your badge. *The New York Times*. Retrieved October 25, 2014, from http://www.nytimes.com/2012/11/04/education/edlife/show-me-your-badge.html

Chambers, M.L. (1994). *Running a perfect BBS*. Indianapolis, IN: Que Publishing.

Chun, W.H.K. (2005). *Control and freedom: Power and paranoia in the age of fiber optics*. Cambridge, MA: The MIT Press.

Clark, A. (2008). *Supersizing the mind: Embodiment, action, and cognitive extension*. New York: Oxford University Press.

Crumlish, C., & Malone, E. (2009). *Designing social interfaces: Principles, patterns, and practices for improving the user experience*. Newton, MA: O'Reilly Media.

Douglass, R.T., Little, M., & Smith, J.W. (2005). *Building online communities with Drupal, phpBB, and WordPress*. New York: Apress.

Duncan, S.C. (2010). Gamers as designers: A framework for investigating design in gaming affinity spaces. *E-Learning and Digital Media*, *7*(1), 21–34.

Fairclough, N. (2003). *Analysing discourse: Textual analysis for social research*. London: Routledge.

Farmer, R., & Glass, B. (2010). *Building web reputation systems: Ratings, reviews & karma to keep your community healthy*. New York: O'Reilly Media.

Figallo, C. (1998). *Hosting web communities: Building relationships, increasing customer loyalty, and maintaining a competitive edge*. New York: John Wiley & Sons.

Fuller, M. (2003). *Behind the blip: Essays on the culture of software*. New York: Autonomedia.

Galloway, A.R. (2006). *Protocol: How control exists after decentralization*. Cambridge, MA: The MIT Press.

Goffman, I. (1967). *Interaction Ritual: Essays on Face-to-Face Behavior*. New York: Doubleday & Company.

Gee, J.P. (2005). *An introduction to discourse analysis: Theory and method* (2nd ed.). London: Routledge.

Gee, J.P. (2010). *How to do discourse analysis: A toolkit*. New York: Routledge.

Gee, J.P., & Hayes, E.R. (2010). *Women and gaming: The Sims and 21st century learning*. Bastingstoke, England: Palgrave Macmillan.

Gibson, W. (1984). *Nuromancer*. New York: Ace.

Greenhow, C., Robelia, B., & Hughes, J.E. (2009). Learning, teaching, and scholarship in a digital age. *Educational Researcher*, *38*(4), 246–259. doi:10.3102/0013189X09336671

Hedtke, J.V. (1990). *Using computer bulletin boards*. Portland, OR: Management Information Source for the Twenty-First Century.

Hine, C. (2000). *Virtual ethnography*. London, England: Sage Publications.

Howard, T. (2010). *Design to thrive: Creating social networks and online communities that last*. Burlington, MA: Morgan Kaufmann.

Hutchins, E. (1995a). How a cockpit remembers its speed. *Cognitive Science*, *19*, 265–288.

Hutchins, E. (1995b). *Cognition in the wild*. Cambridge, MA: The MIT Press.

Ito, M. (2009). *Hanging out, messing around, and geeking out: Kids living and learning with new media*. Cambridge, MA: MIT Press.

Jenkins, H., Purushotma, R., Weigel, M., Clinton, K., & Robison, A. (2009). *Confronting the challenges of participatory culture: Media education for the 21st century*. Chicago, IL: The MacArthur Foundation.

Kim, A.J. (2000). *Community building on the web: Secret strategies for successful online communities*. San Francisco, CA: Peachpit Press.

Kingsley-Hughes, K., & Kingsley-Hughes, A. (2006). *vBulletin: A users guide: Configure, manage and maintain your own vBulletin discussion forum*. New York: Packt Publishing.

Kirk, A. (2007). *Counterculture green: The Whole Earth Catalog and American environmentalism*. Lawrence: University of Kansas Press.

Kozinets, R.V. (2010). *Netnography: Doing ethnographic research online*. London: Sage Publications.

Krug, S. (2000). *Don't make me think: A common sense approach to web usability*. Indianapolis, IN: New Riders Press.

Lankshear, C., & Knobel, M. (2006). *New literacies: Everyday practices and classroom learning* (2nd ed.). Maidenhead, England: Open University Press.

Latour, B. (2005). *Reassembling the social: An introduction to actor-network-theory*. New York: Oxford University Press.

Levy, P. (1997). *Collective intelligence: Mankind's emerging world in cyberspace*. Cambridge, MA: Perseus.

Martin, J.L. (2011). *The explanation of social action*. New York: Oxford University Press.

Maxwell, J.A. (2004). *Qualitative research design: An interactive approach* (2nd ed.). Thousand Oaks, CA: Sage Publications.

Maxwell, J.A. (2011). *A realist approach to qualitative research*. Thousand Oaks, CA: Sage Publications.

Miller, D., & Slater, D. (2001). *The Internet: An ethnographic approach*. Oxford, England: Berg Publishers.

Montfort, N., & Bogost, I. (2009). *Racing the beam: The Atari video computer system*. Cambridge, MA: The MIT Press.

Mytton, D. (2005). *Invision Power Board 2: A user guide: Configure, manage and maintain a copy of Invision Power Board 2 on your own website to power an online discussion forum*. Birmingham, England: Packt Publishing.

O'Keefe, P. (2008). *Managing online forums: Everything you need to know to create and run successful community discussion boards*. New York: AMACOM.

O'Sullivan, M. (2009). Video interview with founder of Vanilla, "A WordPress for Forums." Retrieved October 25, 2014, from: http://vimeo.com/5405609

Owens, T. (2010). Modding the history of science: Values at play in modder discussions of Sid Meier's Civilization. *Simulation & Gaming, 42*(4). doi:10.1177/1046878110366277

Pinch, T.J., & Bijker, W.E. (1984). The social construction of facts and artefacts: Or how the sociology of science and the sociology of technology might benefit each other. *Social Studies of Science, 14*(3), 399–441.

Powazek, D.M. (2002). *Design for community: The art of connecting real people in virtual places*. Indianapolis, IN: New Riders Press.

Powers, M.J. (1997). *How to program a virtual community: Attract new web visitors and get them to stay!* Emeryville, CA: Ziff-Davis.

Rheingold, H. (1993). *The virtual community: Homesteading on the electronic frontier.* Cambridge, MA: The MIT Press.

Robertson, M. (2010, October 6). Can't play, won't play. *Hide & seek: Inventing new kinds of play.* Retrieved October 25, 2014, from http://www.hideandseek.net/2010/10/06/cant-play-wont-play/

Segaran, T. (2007). *Programming collective intelligence: Building smart web 2.0 applications.* Newton, MA: O'Reilly Media.

Shelton, K., & McNeeley, T. (1997). *Virtual communities companion: Your passport to the bold new frontier of cyberspace.* Albany, NY: Coriolis Group Books.

Song, F.W. (2009). *Virtual communities: Bowling alone, online together.* New York: Peter Lang.

Squire, K., & Giovanetto, L. (2008). The higher education of gaming. *E-Learning, 5*(1), 2–28.

Stefanov, S., Rogers, J., & Lothar, M. (2005). *Building online communities with phpBB 2.* New York: Packt Publishing.

Steinkuehler, C., & Chmiel, M. (2006). Fostering scientific habits of mind in the context of online play. In S. Barab, K. Hay, & D. Hickey (Eds.), *Proceedings of the 7th International Conference on Learning Sciences.* Paper presented at the International Society of the Learning Sciences, Bloomington, IN, June 27–July 1 (pp. 723–729). Chicago, IL: International Society of the Learning Sciences.

Steinkuehler, C., & Duncan, S. (2008). Scientific habits of mind in virtual worlds. *Journal of Science Education and Technology, 17,* 530–543. doi:10.1007/s10956-008-9120-8

Trouillot, M. (1995) *Silencing the past: Power and the production of history.* Boston, MA: Beacon Press.

Wolfe, D. (1994). *The BBS construction kit: All the software and expert advice you need to start your own BBS today!* New York: John Wiley & Sons.

Wood, L., & Blankenhorn, D. (1992). *Bulletin board systems for business.* New York: John Wiley & Sons.

Woolgar, S. (1991). *Configuring the user: The case of usability trials.* Cambridge, England: Routledge.

INDEX

A

Actor Network Theory 19–21, 25, 91,
    135

B

Banning Users 30, 38–39, 103, 106–108,
    110, 120, 131
Behaviorist 5, 39 83–86, 122
Bulletin Board Sites (BBS) 57–64, 97–99,
    115

C

Censor 69, 105–106, 110
Cognition 22–23, 25, 123
Collective Intelligence 25–27, 122
Community Guidelines 69, 104–107, 120

Community Manager

Community Manager 1, 5, 9, 11, 13, 20, 30,
    33–34, 36, 82, 86, 88–91, 94, 101, 129
Configure/Configuration 8, 11–12 18, 22,
    32, 34, 58, 61, 73, 92, 94, 97, 108,
    122, 124
Cyberspace 9, 40, 63–67, 70, 72–75

D

Database 17, 46, 64, 74–75, 99, 101,
    112–113, 118
Digital Records 123–129
Distributed Intelligence/Distributed
    Cognition 24–25

E

Educators 2, 5, 8, 125, 126
Email 58, 96, 100–101, 108, 112

F

Facebook 35, 77–80
Flickr 77, 116–118

G

Gameification 84
Google 7, 11, 23–25, 52, 79, 98, 100, 122
Governance 47, 52, 55–56, 69–70, 86, 89, 104
Gratification 83–86

H

Hierarchy 30, 85, 86, 123
Hosting 64, 69–72, 96, 134
HTTP 10, 18, 21, 97, 103, 108, 109

I

Identity 103
Ideology 2, 4, 8, 16, 30–31, 47, 53, 55, 63, 68, 73, 119, 122–123, 131–132
Information architecture 29, 47, 81, 92, 96, 120, 131

K

Knowledge Base 7, 10, 26, 122

L

LinkedIn 79

M

Marketing 31, 34, 36–37, 51, 74, 83, 97, 103

Mediation 1, 8–9, 16–17, 23, 53, 83, 105, 118, 121–123, 129
Motivation 8, 17, 21, 30, 33, 39, 75, 83, 85, 117, 126

P

phpBB 73–80, 97–105
Platform Studies 17
Pruning Discussions 101–102

R

Reputation systems/User Ranks 16, 21, 75–76, 80–86, 111–118
Rules/Norms 17, 19–20,37, 41, 55, 61–62, 69–70, 89, 97, 103–106, 120, 131

S

Screen 8–9, 18, 65, 75, 93, 96, 99, 102, 120, 126
Site Visitors 50–52, 63–64, 70, 74, 96, 99
Social Construction of Technology 19–21, 135
Social Network 77–82
Sysop 11, 57–62, 90, 108, 113, 133

T

Tagline 68, 92, 126
TCP/IP & DNS 10, 20, 120
Twitter 77, 78, 80, 117, 118

U

Ubuntu 86, 89
User Agency 38, 52, 80, 82, 90–92, 102, 113

V

Vanilla Forums 7, 16–17, 21, 127, 135
Vbulletin 73, 89, 92, 97–103

W

Web Forum 7–26, 73–80, 103–112, 111
World Wide Web 10, 13, 25, 35, 50–69,
      65, 88, 122

Y

Yelp 77, 122

Z

Zotero 15–16, 21, 39, 46

Colin Lankshear & Michele Knobel

General Editors

New literacies emerge and evolve apace as people from all
walks of life engage with new technologies, shifting values
and institutional change, and increasingly assume 'postmod-
ern' orientations toward their everyday worlds. Despite many
efforts to take account of such changes, educational institutions largely remain out of touch with the range of new
ways of making and sharing meanings that increasingly mediate and shape the lives of the young people they teach and
the futures they face. This series aims to explore some key
dimensions of the changes occurring within social practices
of literacy and the educational challenges they present,
with a view to informing educational practice in helpful
ways. It asks what are new literacies, how do they impact on
life in schools, homes, communities, workplaces, sites of
leisure, and other key settings of human cultural engagement, and what significance do new literacies have for how
people learn and how they understand and construct knowledge. It aims to challenge established and 'official' ways
of framing literacy, and to ask what it means for literacies
to be powerful, effective, and enabling under current and
foreseeable conditions. Collectively, the works in this series will help to reorient literacy debates and literacy
education agendas.

For further information about the series and submitting
manuscripts, please contact:

Michele Knobel & Colin Lankshear
Montclair State University
Dept. of Education and Human Services
3173 University Hall
Montclair, NJ 07043
michele@coatepec.net

To order other books in this series, please contact our
Customer Service Department at:
(800) 770-LANG (within the U.S.)
(212) 647-7706 (outside the U.S.)
(212) 647-7707 FAX

Or browse online by series at:
www.peterlang.com

ADVANCE PRAISE FOR designing online communities

"*Designing Online Communities* is a must-have for anyone designing or researching online communities, particularly for learning. Trevor Owens's work is both comprehensive and eminently readable, a sweeping look at the technologies, design patterns, and cultural forms they produce that is both theoretically ambitious and grounded in examples and tools that will help you develop, research, and manage online communities."

—Kurt Squire, University of Wisconsin

"Part enabler, part denier, full-on technological mediation, web forums offer a fascinating entry point into the interplay of software and social interaction. In *Designing Online Communities*, Owens deftly mixes actor-network theory, discourse analysis, and other approaches, writing with clear language and insight to expose the ideologies inherent in seemingly pedestrian historical artifacts—how-to books for web forum administrators. His engaging analysis gives clarity to how the design strategies implicit in code influence the ways we build conversations, relationships, and communities on the web."

—Jefferson Bailey, Internet Archive

"Can media archaeology have a methodology? Does software studies need data sets? In *Designing Online Communities*, Owens presents a bracing case study that not only contributes to our understanding of lives lived online, but also joins the empirical rigor of applied social science with leading-edge digital and media studies."

—Matthew Kirschenbaum, University of Maryland

"An important read for educators interested in using and building online communities. Owens asks us to consider how technologies reflect and shape permissions and control, and how the managers and builders of online communities wield power beyond simply an offer of 'connectivity.'"

—Audrey Watters, Hack Education

"At a time when online communities are ubiquitous, and in some cases larger than most countries, it is critical that we understand how they are composed—technologically, psychologically, and sociologically. Owens shrewdly looks back to early bulletin boards and web forums to grasp the nature of these modern communities, how they arose, how they dealt with bad behavior and the inevitable disagreements between members, and how all of this was represented in rhetoric and code. This book provides essential context for our shared online existence."

—Dan Cohen, Digital Public Library of America